# COCKPIT AUTOMATION
## For General Aviators and Future Airline Pilots

# COCKPIT AUTOMATION

## For General Aviators and Future Airline Pilots

STEPHEN M. CASNER

ILLUSTRATED BY DOUGLAS A. DUPUIE

Iowa State Press
A Blackwell Publishing Company

**STEPHEN M. CASNER**, Ph.D., is a research scientist at NASA's Ames Research Center in California. Steve holds an Airline Transport Pilot certificate with type ratings in the Boeing 737 and Cessna Citation. Steve is also a Gold Seal Certified Flight Instructor.

**DOUGLAS A. DUPUIE** is currently a First Officer on the Canadair Regional Jet. Doug holds a B.A. in psychology from Metropolitan State College of Denver. Doug is also a Gold Seal Certified Flight Instructor.

The purpose of this book is to provide information on airplane operations. The user of this information assumes all risk and liability arising from such use. Neither Iowa State Press nor the author can take responsibility for the actual operation of an aircraft or the safety of its occupants.

© 2002 Iowa State Press
A Blackwell Publishing Company
All rights reserved

Iowa State Press
2121 State Avenue, Ames, Iowa 50014

Orders:    1-800-862-6657
Office:    1-515-292-0140
Fax:       1-515-292-3348
Web site:  www.iowastatepress.com

Authorization to photocopy items for internal or personal use, or the internal or personal use of specific clients, is granted by Iowa State Press, provided that the base fee of $0.10 per copy is paid directly to the Copyright Clearance Center, 222 Rosewood Drive, Danvers, MA 01970. For those organizations that have been granted a photocopy license by CCC, a separate system of payments has been arranged. The fee code for users of the Transactional Reporting Service is 0-8138-2300-5/2002 $0.10.

∞ Printed on acid-free paper in the United States of America

First edition, 2002

**Library of Congress Cataloging-in-Publication Data**
Casner, Stephen M.
    Cockpit automation: for general aviators and future airline pilots / Stephen M. Casner; Douglas A Dupuie, illustrations.— 1st ed.
        p. cm.
    Includes bibliographical references and index.
    ISBN 0-8138-2300-5
    1. Airplanes—Automatic control. 2. Avionics. 3. Automatic pilot (Airplanes). 4. Aids to air navigation. I. Title.
TL589.4 .C37 2002
629.135—dc21                                                    2001005967

Last digit is the print number: 9 8 7 6 5 4 3 2 1

# CONTENTS

    **List of Figures and Captions,** vi

    **Acknowledgments,** ix

**1**   Getting Started, 3

**2**   Planning the Flight Route, 9

**3**   Following the Flight Route, 39

**4**   Modifying the Flight Route, 61

**5**   Advanced Maneuvers, 73

**6**   Flying with an Autopilot, 101

**7**   Avoiding Traffic, Terrain, and Weather, 129

**8**   The Human Factor, 151

**9**   Proficiency Standards, 157

**10**  For Airline-Bound Pilots, 159

    **Glossary,** 167

    **Index,** 173

# LIST OF FIGURES AND CAPTIONS

**Chapter 1**
Figure 1-1: Cockpit automation systems found in the modern small-airplane cockpit.

**Chapter 2**
Figure 2-1: A conventional flight plan.
Figure 2-2: Global positioning system (GPS) computers.
Figure 2-3: Documents that certify a GPS unit for IFR operations.
Figure 2-4: Checking navigation database currency.
Figure 2-5: The keyboard and monitor on the GPS computer.
Figure 2-6: Pages.
Figure 2-7: Switching between pages.
Figure 2-8: Extended pages.
Figure 2-9: Using cursor mode to enter data.
Figure 2-10: The flight plan page.
Figure 2-11: Activating a stored flight plan.
Figure 2-12: Building a flight plan from scratch.
Figure 2-13: Selecting a departure procedure.
Figure 2-14: Selecting a runway and departure transition.
Figure 2-15: Loading your selections into the flight plan.
Figure 2-16: Selecting and loading an approach.
Figure 2-17: Reviewing the flight route.
Figure 2-18: Checking the desired tracks between waypoints.
Figure 2-19: A great circle route.

**Chapter 3**
Figure 3-1: RAIM alerts.
Figure 3-2: The flight plan page and the active waypoint.
Figure 3-3: The navigation page.
Figure 3-4: Engaging the sequencing mode.
Figure 3-5: Waypoint alerting.
Figure 3-6: Turn anticipation.
Figure 3-7: Waypoint sequencing.
Figure 3-8: A navigation indicator slaved to the GPS computer.

# LIST OF FIGURES AND CAPTIONS

Figure 3-9: Planning a descent.
Figure 3-10: Planning a descent with the computer.
Figure 3-11: How winds cause you to drift off the planned descent path.
Figure 3-12: Monitoring progress in the descent.
Figure 3-13: Top-of-descent alerts.
Figure 3-14: Terminal mode.
Figure 3-15: Approach mode.

## Chapter 4
Figure 4-1: The Direct-To function.
Figure 4-2: Adding waypoints to your route.
Figure 4-3: Deleting waypoints from your route.
Figure 4-4: Selecting a different approach or transition.
Figure 4-5: Emergency diversions.

## Chapter 5
Figure 5-1: Sunol Five departure.
Figure 5-2: Need to track a different course to the active waypoint.
Figure 5-3: Using OBS mode to intercept a different course to the active waypoint.
Figure 5-4: Need to change the active waypoint.
Figure 5-5: Using the Direct-To function to change the active waypoint.
Figure 5-6: Changing the course to the active waypoint.
Figure 5-7: The vector-to-final feature.
Figure 5-8: Flying a holding pattern with the nonsequencing mode.
Figure 5-9: An approach with a procedure turn.
Figure 5-10: Flying a procedure turn with the nonsequencing mode.
Figure 5-11: Flying a missed approach procedure.
Figure 5-12: Computers sometimes make mistakes.
Figure 5-13: Pilots sometimes make mistakes.

## Chapter 6
Figure 6-1: Engaging the autopilot.
Figure 6-2: A flight director presented on an electronic flight instrument system (EFIS) display.
Figure 6-3: The vertical speed function.
Figure 6-4: Autopilot trim commands.
Figure 6-5: The altitude capture function.
Figure 6-6: The heading function.
Figure 6-7: Tracking a VOR radial using the navigation function.
Figure 6-8: Intercepting and tracking a VOR radial.
Figure 6-9: Flying an assigned heading to intercept a VOR radial.
Figure 6-10: The approach function.
Figure 6-11: Flying a localizer approach with the approach function.
Figure 6-12: Flying an assigned heading to intercept an approach course.
Figure 6-13: Flying a precision approach with the approach and glide slope functions.
Figure 6-14: Disconnecting the autopilot.
Figure 6-15: Mode confusion.

Figure 6-16: A similar autopilot made by a different manufacturer.
Figure 6-17: A more sophisticated autopilot.

**Chapter 7**
Figure 7-1: TCAS traffic displays.
Figure 7-2: Traffic display symbology.
Figure 7-3: TCAS control panel.
Figure 7-4: TCAS II control panel.
Figure 7-5: Integrated TCAS and vertical speed indicator.
Figure 7-6: A more sophisticated integrated TCAS and vertical speed indicator.
Figure 7-7: Terrain presented on a multifunction display (MFD).
Figure 7-8: Terrain display symbology.
Figure 7-9: Airborne weather radar system.
Figure 7-10: Weather radar display symbology.
Figure 7-11: Sharp intensity gradients suggest severe turbulence.
Figure 7-12: Cell shapes that suggest the presence of hail.
Figure 7-13: A weather radar beam.
Figure 7-14: Weather radar antenna set too low.
Figure 7-15: Weather radar antenna set too high.
Figure 7-16: Radar attenuation.
Figure 7-17: Importance of range settings on the weather radar display.
Figure 7-18: Lightning detection system displays.

**Chapter 8**
Figure 8-1: A small error can sometimes have big consequences.

**Chapter 10**
Figure 10-1: *The Pilot's Guide to the Modern Airline Cockpit.*
Figure 10-2: The flight management computer's control display unit (CDU).
Figure 10-3: The navigation display (ND).
Figure 10-4: The autopilot's mode control panel (MCP).
Figure 10-5: The flight director on the primary flight display (PFD).

# ACKNOWLEDGMENTS

I WOULD LIKE TO THANK THE pilots that flew with me and provided valuable insights about cockpit automation as these materials were being developed and tested: Alex Alugas, Aric Arnold, Kristine Anderson, Rose Ashford, Immanuel Barshi, Ron Bawden, Sean Belcher, Key Dismukes, Deborah Dunn, Doug Dupuie, Mike Feary, Amy Flores, Daniel Fout, Ed Hutchins, Homi Irani, Zach Larner, Giorgio Magnoni, Lindsay Miller, Eric Norgaard, Ray Oyung, Silvia Peschke, Boris Ricci, Stephan Romahn, John Shopland, Dave Stamps, and Eric Villeda.

Bel-Air International in San Carlos, California provided a welcome flight school setting in which to develop and test these materials with student pilots and flight instructors. Terry McKenna kept the airplane in top shape. Sierra Academy of Aeronautics also graciously allowed me to recruit their students and flight instructors for the project. Additional airplanes were provided by Palo Alto Flying Club.

Mike Lewis, Kevin Jordan, Tina Beard, Brian Smith, and Key Dismukes supported the project from start to finish. Immanuel Barshi, Mike Feary, and Ed Hutchins provided helpful comments on the manuscript. I thank many other colleagues for countless discussions about cockpit automation: Everett Palmer, Earl Wiener, Pete Polson, Lance Sherry, Reid Fairburn, Randy Mumaw, Ed Hutchins, and Bill Bulfer to name just a few. Gavin Tanchuck, Cliff Bonner, Brennan Fallon and a lot of student pilots taught me a lot about what it means to be a good flight instructor: something I'm trying to improve upon all the time.

# COCKPIT AUTOMATION

## For General Aviators and Future Airline Pilots

# CHAPTER 1

# Getting Started

*This chapter gets you started in the exciting world of cockpit automation: the high-tech computers and displays that are now common in the modern small airplane cockpit. You will find out why learning about cockpit automation is a good idea, both for the general aviation pilot as well as for the pilot with aspirations to fly for an airline. You will learn how to use this book, together with a collection of free computer-based simulators, to master the basic skills you will need to confidently act as pilot-in-command in the modern high-technology cockpit.*

This book introduces you to a new generation of airplane that features **cockpit automation:** the collection of computers and displays that have taken the place of the traditional flight instruments, navigation, and control systems found in older-generation airplanes. These high-technology cockpit systems have traditionally been found only in the modern airline cockpit, but the pilots of smaller airplanes can today enjoy much of the excitement, benefits, and challenges offered by such systems. Almost every small airplane manufactured today features at least some of the same kinds of computers-in-the-cockpit found in the big airliners.

Small airplanes today feature, at the least, a **global positioning system (GPS) computer** that can assist you in performing the traditional tasks of flight planning and navigation. Some airplanes also feature **autopilots** and **flight director systems** that can assist you when steering the airplane along the route that you have planned. More advanced airplanes offer **ground proximity warning systems (GPWS), traffic collision avoidance systems (TCAS), electronic flight instrument systems (EFIS),** and **airborne weather detection systems.**

These systems, shown in Figure 1–1, offer a powerful set of cockpit resources to the small-plane pilot. Whether you are an avid general aviator, a corporate pilot, or an aspiring airline pilot, it is hoped that you will use this book to acquaint yourself with these systems and begin to experience the changing job of flying in the modern high-tech cockpit.

# 4 COCKPIT AUTOMATION for General Aviators and Future Airline Pilots

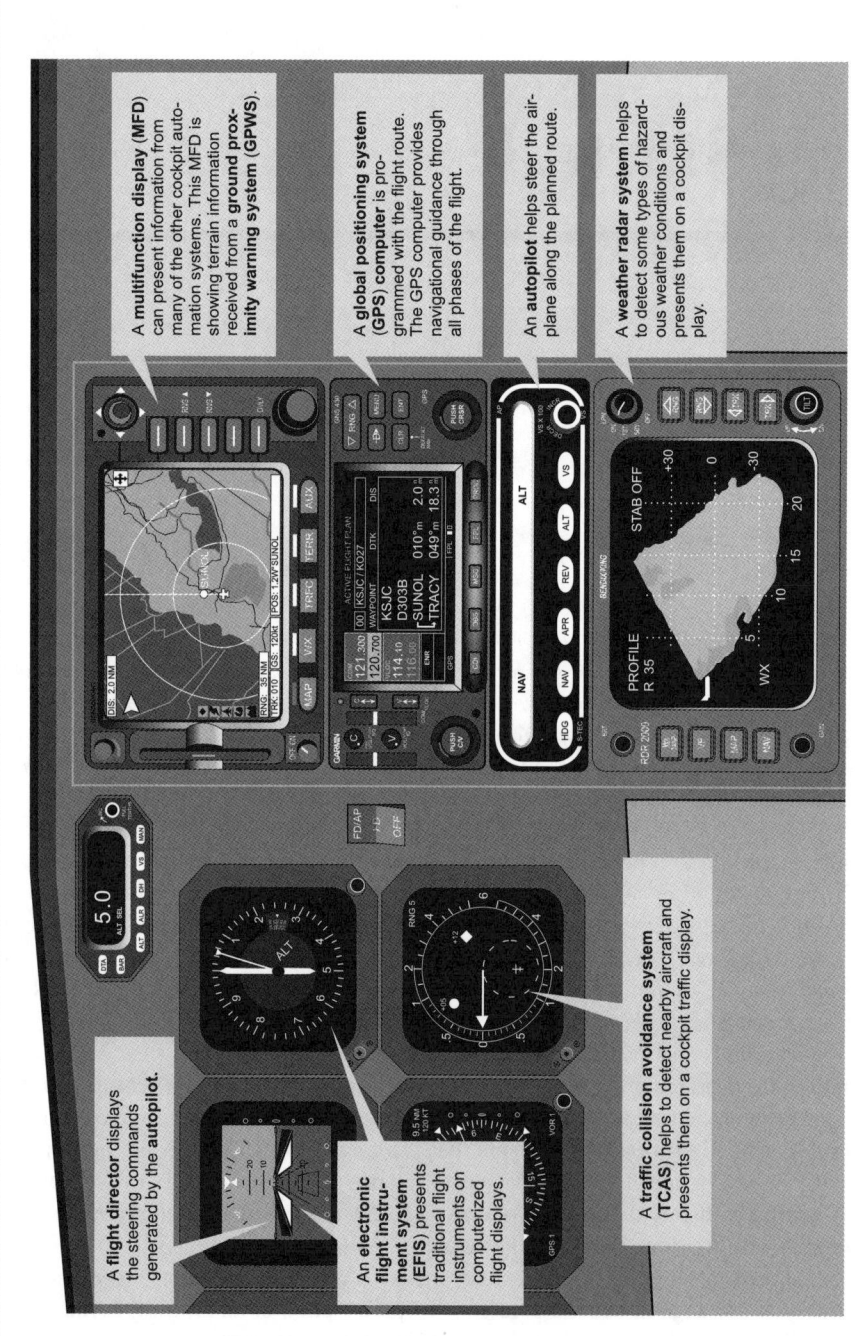

1.1. Cockpit automation systems found in the modern small-airplane cockpit.

# 1 / Getting Started

## WHY LEARN ABOUT COCKPIT AUTOMATION?

For the general aviation pilot, the most motivating reason to learn about cockpit automation is that practically every new airplane manufactured today contains some type of cockpit automation as standard equipment. Too often, pilots not familiar with these new technologies leave the equipment turned off in favor of more familiar equipment and methods. The problem with this scenario is that, aside from missing an exciting learning opportunity, the pilot has new cockpit automation technologies right at hand that can significantly enhance both safety and productivity.

For the career-minded pilot, taking advantage of the cockpit automation now found in small training airplanes represents a chance to get a head start on facing the inevitable challenge of walking into a modern airline cockpit. The airplanes now operated by regional airlines are fully equipped with the latest in cockpit automation technology. These systems represent a significant learning challenge for new-hire pilots, especially those coming directly from single-engine training airplanes with no cockpit automation experience.

## ISN'T LEARNING ABOUT COCKPIT AUTOMATION MOSTLY ABOUT KNOWING WHAT BUTTONS TO PUSH?

The most interesting thing I discovered when developing and testing the materials presented in this book is that becoming a proficient cockpit automation user means more than learning how to push buttons and twist knobs. In fact, learning to operate the knobs, dials, and displays was the least difficult challenge for pilots new to the systems we studied. After working with scores of student pilots and shooting hundreds of GPS approaches, I found that most students learned the procedures quickly and recalled them easily when they were regularly used in practice.

What presented the biggest challenge to most students was understanding and maintaining an awareness of what the automation was currently configured to do, especially in challenging scenarios presented by air traffic control (ATC). Although students had little trouble understanding these new skills and concepts in the classroom, they proved difficult for students to incorporate into their existing habit patterns when flying the airplane. Students had to be constantly reminded that the automation had a plan of its own, and that this plan had to be continually monitored by the student. As one expert has pointed out, the most common three questions asked in the modern automated cockpit are as follows:

1. Why did it do that?
2. What's it doing now?
3. What's it going to do next?

You will likely find that developing the skills required to work as a team with cockpit automation will challenge you the most.

## HOW TO PRACTICE WHAT YOU LEARN

The chapters of this book take you through many of the important topics on your road to cockpit automation proficiency. Most chapters conclude with a practice session designed to allow you to practice what you have learned from reading. The practice sessions allow you to act as your own flight instructor and to accomplish the exercises using either a simulator or an aircraft. The book assumes the use of at least an IFR-capable GPS computer. The book contains descriptions and illustrations for the Bendix/King KLN 94 and the Garmin GNS 430 GPS computers. The manufacturers of these computers now offer free downloadable PC-based simulators for these products. You can download these simulators now at

www.garmin.com
www.bendixking.com

You will also want to download the manufacturers' pilot information manuals for these products while you are there.

If you have a different GPS computer in mind, check my Web site where you will find supplements to this book for several other GPS computers:

www.iowastatepress.com/casner/

The chapter about autopilots can be understood without the use of an autopilot, although experience with an autopilot provides valuable insight into the nature of highly automated flight. The book contains descriptions and illustrations for the S-Tec System 55 autopilot. S-Tec offers a free downloadable training module for the System 55 autopilot, as well as pilot information manuals for their products on their Web site:

www.s-tec.com

Several other autopilots are briefly discussed, and the manufacturer's information manuals can be downloaded from their Web site:

www.bendixking.com

This book will provide you with a set of concepts and skills that you will need to safely and proficiently use cockpit automation. As you work your way through the book, it is strongly recommended that you stick to the basics. Although you will encounter many details that are not explained here, write down your questions and save them for later. Your goal is not to understand every symbol that appears on every display of Company X's equipment. Rather, your goal is to master the basic concepts and habit patterns that will prepare you for using cockpit automation in real flight situations. You are encouraged to obtain a copy of the manufacturer's operating manual to help answer your questions about the details of the equipment you are using. Although not designed as primary training materials, the manufacturers' manuals are your most important resource for reference information about technical details of the equipment you are using. Operating manuals for other manufacturers' equipment discussed in the book are available on their Web sites:

www.bfgavionics.com
www.insightavionics.com

It is also recommended that you read Section 1-1-21 of the current edition of the *Aeronautical Information Manual* (AIM).

# 1 / Getting Started

If you are using an airplane for the practice sessions, plan to take a safety pilot with you during the day and under VFR conditions. Plan to make plenty of mistakes and learn many things, even after you have read the book. If you are a Certified Flight Instructor (CFI) and are teaching cockpit automation to a student, the chapters in the book should serve well as the basis for your lesson plans.

In the back of this book, you will find a DVD (digital video disc). This disc provides a video demonstration of each of the skills covered in this book. The DVD is not intended for use as your primary source of instruction. View the demonstrations on the DVD only after you have read each chapter and feel that you understand the concepts and skills well enough to try them yourself. The DVD will also serve to refresh your memory after you have learned the skills and concepts but have been away from them for some time.

## WHERE TO GO FROM HERE

If you are a general aviator, plan to learn many things from your experiences with the cockpit automation systems described in this book. Keep in mind that reading a book and taking a practice flight or two does not make one an expert. Even after accumulating over 1,500 flight hours using an IFR GPS computer, I learn something new almost every time I fly with computers in my cockpit.

If you are preparing for a career as an airline pilot, the next step in learning about cockpit automation is to master the cockpit automation systems found in modern jet transport airplanes. Designed as a follow-up to this book, *The Pilot's Guide to the Modern Airline Cockpit,* published by Iowa State University Press, starts where this book leaves off and takes you to that final frontier of cockpit automation.

Fly safely and enjoy the ride.

# CHAPTER 2

# Planning the Flight Route

*This chapter teaches you how to plan and prepare for an IFR flight in a modern automation-equipped aircraft. You will learn how to work together with a device called a global positioning system (GPS) computer to plan the flight. You will see how pilots perform the tasks that require decision making and judgment and how the GPS computer can help with some of the tedious math. By the end of this chapter, you will have begun to learn your way around a GPS computer and be able to use the GPS computer to exercise three important flight-planning skills:*

> *SKILL 1: Determine if your GPS computer is approved for IFR navigation.*
> *SKILL 2: Program the GPS computer with a flight plan.*
> *SKILL 3: Check the flight plan to ensure that you and the computer have the same plan.*

Your introduction to cockpit automation begins with the first step of any flight: flight planning. Planning a flight in a modern, automated cockpit isn't much different than flight planning in an older-generation airplane. You still have to check the weather, consult your charts, and choose a suitable route including airways, terminal procedures, and alternate airports. You then make the familiar list of checkpoints along your route; measure the distance between the checkpoints; and determine the courses, time, and fuel required to make your way between them. This process should sound quite familiar by now, and the good news is that you are probably already an expert at the job of preparing flight plans such as the one shown in Figure 2-1.

What is different in the automated cockpit is that you now have a computer onboard that will help you in many ways as you plan and make your way along your route. In the modern small-airplane cockpit, you get this automated assistance from a **global positioning system (GPS) computer.** A GPS computer allows you to enter the details of the planned flight route and then provides many interesting services as you fly along the route. When you enter the details of your flight route, the GPS computer does all of the same mathematical calculations (e.g., course, time, and distance) that you do, and it allows you to compare your work with the results produced by the computer.

## NAVIGATION LOG

| Aircraft Number: N4334V | Dep: KSJC | Dest: O27 | Date: 11/06/02 |
|---|---|---|---|

| Clearance: |
|---|
| C - O27 |
| R - SUNOL FIVE ECA, Direct O27 |
| A - CLB 5000 |
| F - 121.3 |
| T - 1719 |

Estimated Time en Route = 0:37

| Check Points (Fixes) | Ident | Course (Route) | Altitude | Mag Crs. | Fuel | Dist. | GS | Time Off |
|---|---|---|---|---|---|---|---|---|
| | Freq. | | | | Leg / Rem. | Leg / Rem. | Est. / Act. | ETE / ATE | ETA / ATA |
| KSJC  TWR 120.7 | DEP  121.3 | | | | 48 | 67.8 | | |
| SJC  1.8 DME | SJC  114.1 | 303 | CLB | | 1.8  66 | 8  58 | 81 | 0:01 |
| SUNOL | SJC  114.1 | 009 | 5000 | | 1.8  46.2 | 8  58 | 81 | 0:05 |
| TRACY | ECA  116.0 | 049 | 5000 | | 1.4  44.8 | 18  40 | 95 | 00:11 |
| ECA (IAF) | ECA  116.0 | 049 | 3000 | | 1.0  43.8 | 15  25 | 114 | 0:08 |
| MOTER | ECA  116.0 | 089 | 2000 | | 0.5  43.3 | 8  17 | 103 | 0:04 |
| ELTRO (FAF)  D 13.5 ECA | ECA  116.0 | 089 | 2000 | | 0.4  42.9 | 5.5  11.5 | 90 | 0:03 |
| MAP D 18.0  RW 10 | ECA  116.0 | 089 | MDA 720 | | 0.3  42.6 | 4.5  7 | 90 | 0:02 |
| WRAPS  D 16.7 LIN | LIN  114.8 | 304 | 3000 | | 0.4  42.2 | 7  0 | 90 | 0:03 |

**Figure 2.1. A conventional flight plan.**

In this book, we will work with two different GPS computers shown in Figure 2-2: the Garmin GNS 430 and the Bendix/King KLN 94. If you have Internet access, take the time to download the free PC-based simulators made available on the manufacturers' Web sites listed in Figure 2-2.

### SKILL 1: DETERMINE IF YOUR GPS COMPUTER IS APPROVED FOR IFR NAVIGATION

Before using any GPS computer for an IFR flight, you must first make sure that the unit you are using is approved for IFR navigation. Only specific GPS computers are approved for IFR navigation and it is important to make this determination before flying with any particular equipment. No

# 2 / Planning the Flight Route

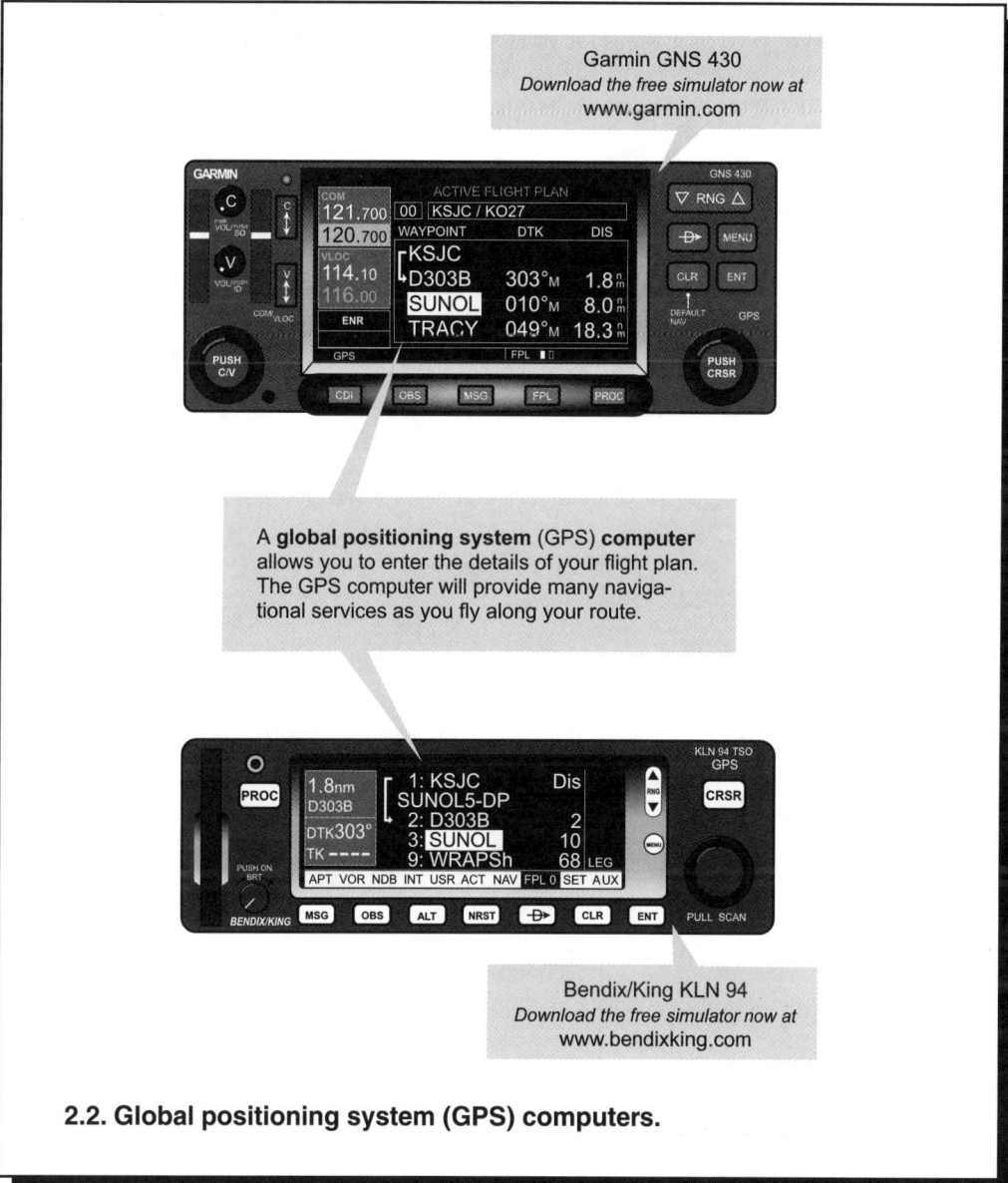

**2.2. Global positioning system (GPS) computers.**

handheld GPS computer is approved for IFR navigation, and many panel-mounted computers are restricted to VFR use only. Even if your GPS computer is IFR capable, it is possible to install the computer in an airplane without having gone through the IFR certification process. Such a unit is not approved for IFR navigation.

## POH Supplement and Form 337

The first place to check, when determining IFR certification for a GPS computer, is the pilot operating handbook (POH) or airplane flight manual (AFM). For every aircraft with an approved IFR GPS computer, the POH should contain a supplement, signed by an FAA inspector, explicitly stating that the unit has been approved for IFR navigation. The POH supplement contains the official operating procedures and limitations for your computer, so it is a good idea to familiarize yourself with that document. An FAA Form 337 approves the installation of the GPS computer in the airplane and officially returns the airplane to service. A sample POH supplement and Form 337 are shown in Figure 2-3.

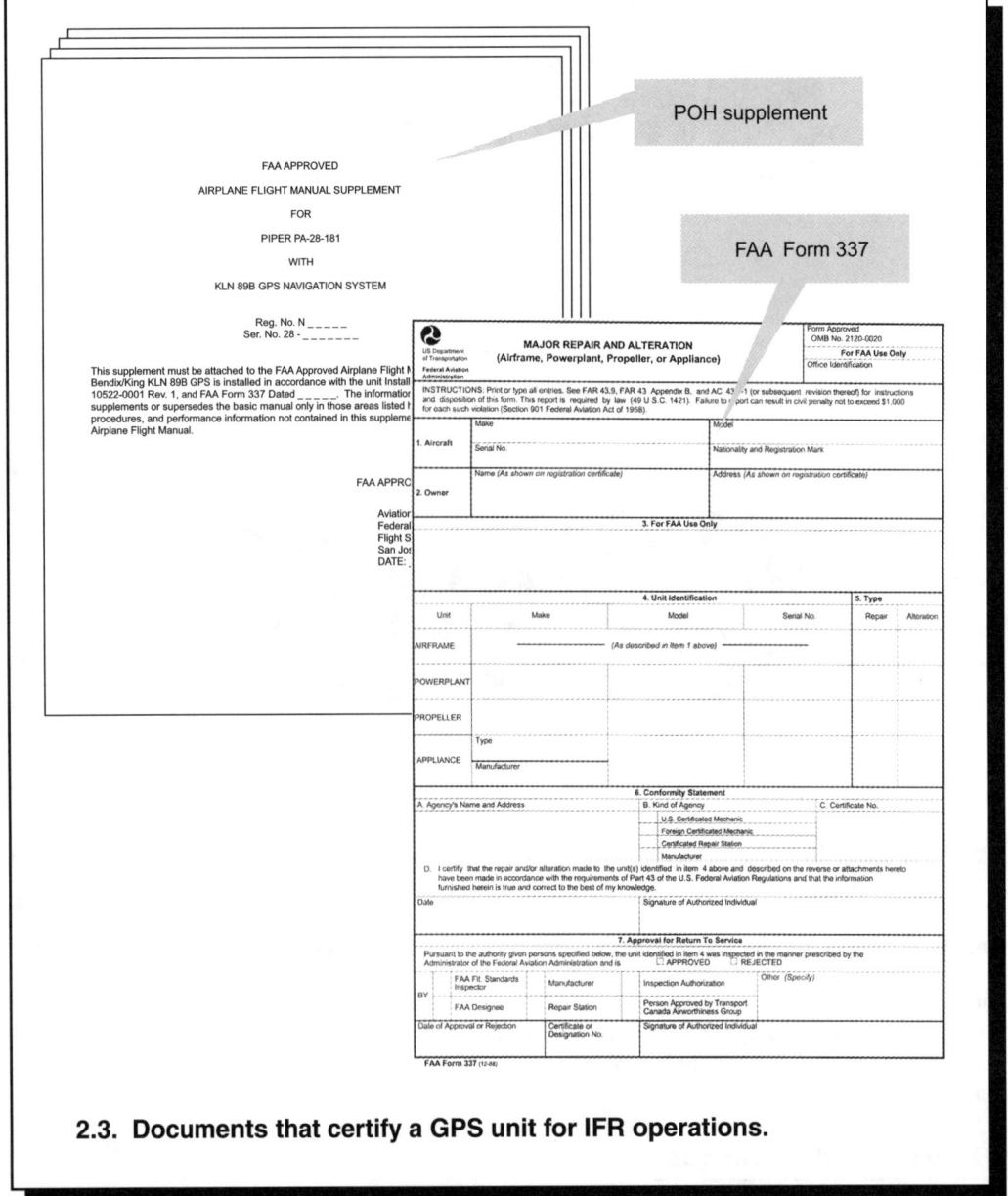

2.3. Documents that certify a GPS unit for IFR operations.

## 2 / Planning the Flight Route

### Current Navigation Database

Every IFR-capable GPS computer makes use of an electronic **navigation database** that contains all of the information found on your en route and terminal procedure charts. As with paper charts, it is a requirement that this electronic navigation database be current. Every GPS computer gives frequent alerts to the user when the navigation database is out of currency. These alerts occur when the unit is powered up and at important times during the operator's interactions with the unit. Figure 2-4 shows the startup screens for the two GPS computers we discuss throughout the book. **Note:** Chapters 2, 3, 4, and 5 present two versions of each figure mentioned in the text: one for each of the two GPS computers being discussed. If you are using a Garmin GNS 430 computer, use the figure that contains the Garmin icon in the lower right corner of each figure. If you are using a Bendix King KLN 94, use the figure with the Bendix King icon. We have also set up a web site (www.iowastate.com/casncr/) that contains figures for several other GPS computers, such as the Garmin GNS 530 and the Bendix King KLN 89B.

### Alternate Means of Navigation

To be approved for IFR flight, an aircraft also has to be equipped with an approved and alternate means of navigation appropriate to the flight. A VOR receiver is a popular choice of approved alternate equipment. Make sure this equipment is onboard the aircraft, is operational, and all required checks have been performed (e.g., 30-day VOR check).

## SKILL 2: PROGRAM THE GPS COMPUTER WITH A FLIGHT PLAN

Since you are going to use the GPS computer as your principal means of navigation, you want to provide the computer with the details of your flight plan. You never want to take off with only a part

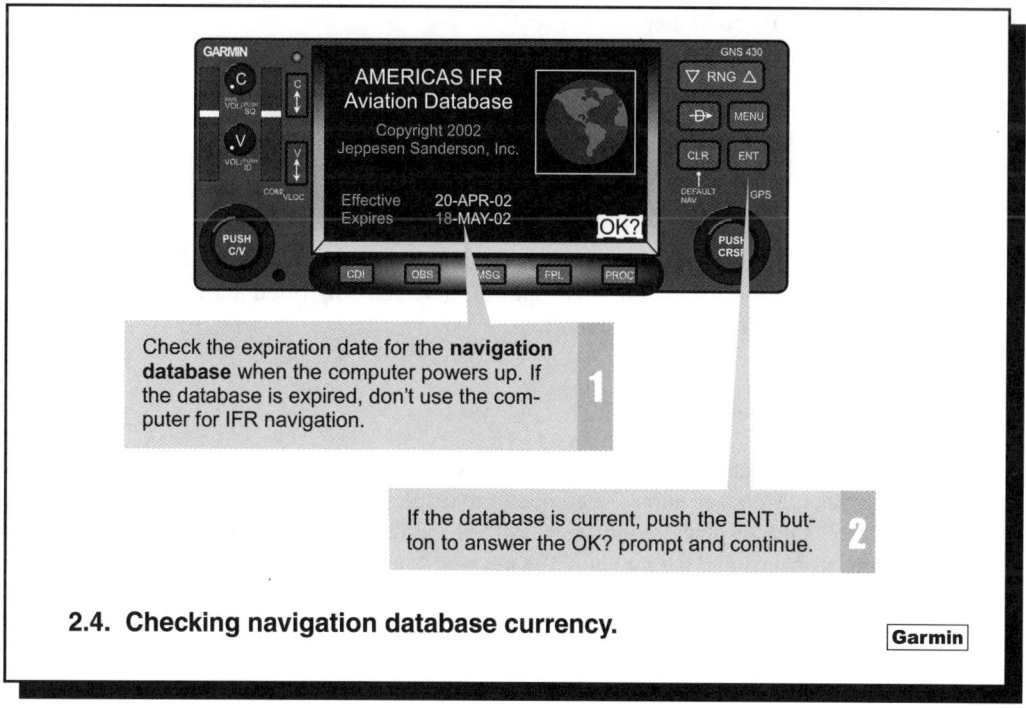

2.4. Checking navigation database currency.

**14**      COCKPIT AUTOMATION for General Aviators and Future Airline Pilots

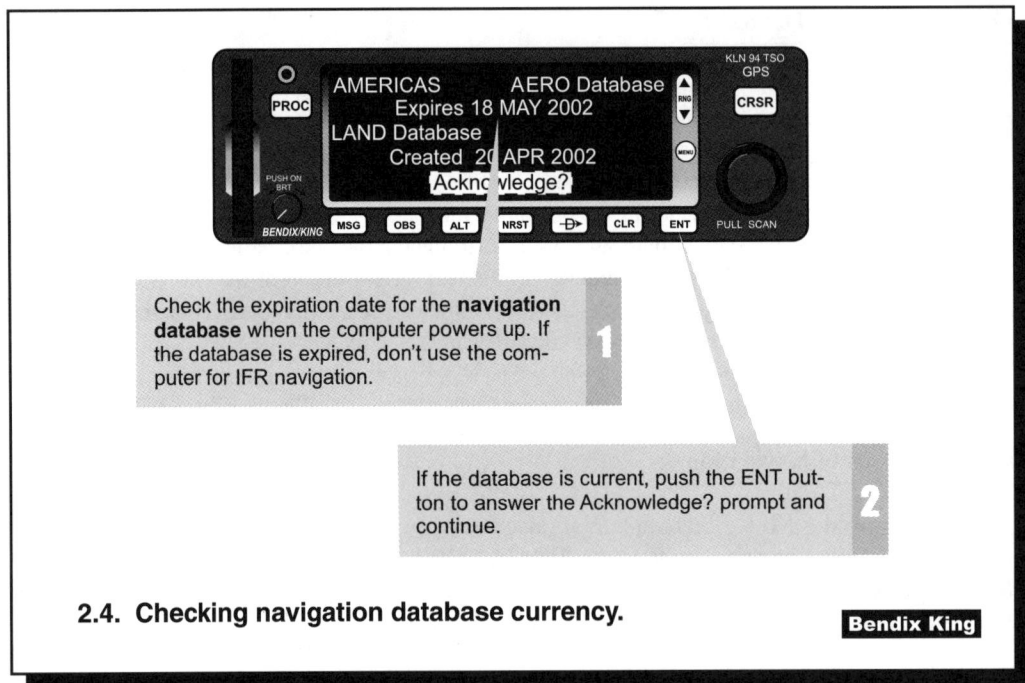

2.4. Checking navigation database currency.

of the flight plan in the computer, with the intention to put the rest in later. It's much easier to program a computer when you are on the ground than it is when flying in the clouds. So your first step is to enter the details of the flight plan you have created using your paper charts and flight plan log.

Before you start twisting knobs and dials, take a moment to familiarize yourself with what you see on the front of the GPS computer.

### Learning Your Way Around the GPS Computer

Let's begin with some simple concepts that will help you understand how modern flight computers organize the volumes of information contained in them, and how they present this information to you so that you can view, use, and manipulate it. Remember, the goal is not to simply memorize the procedures for operating the computer but rather to understand and reason your way through the procedures. You'll find that this depth of understanding often allows you to re-create procedures when you have forgotten a step or two.

#### *Computer, Keyboard and Monitor*

Most every computer has a **keyboard** and a **monitor.** Keyboards allow you to input information into the computer. Monitors allow the computer to display information to you. The monitor for any GPS computer is the display area in the middle of the unit. The keyboard is the collection of knobs and buttons that surround the monitor. The keyboards and monitors for the two GPS computers are illustrated in Figure 2-5.

## 2 / Planning the Flight Route

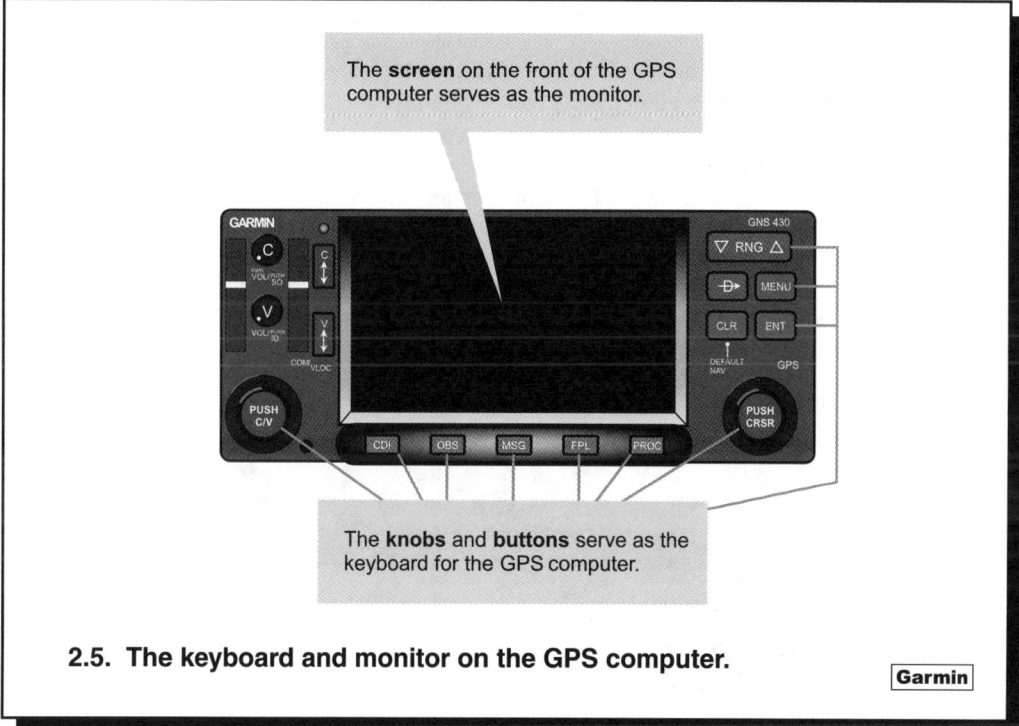

2.5. The keyboard and monitor on the GPS computer.  Garmin

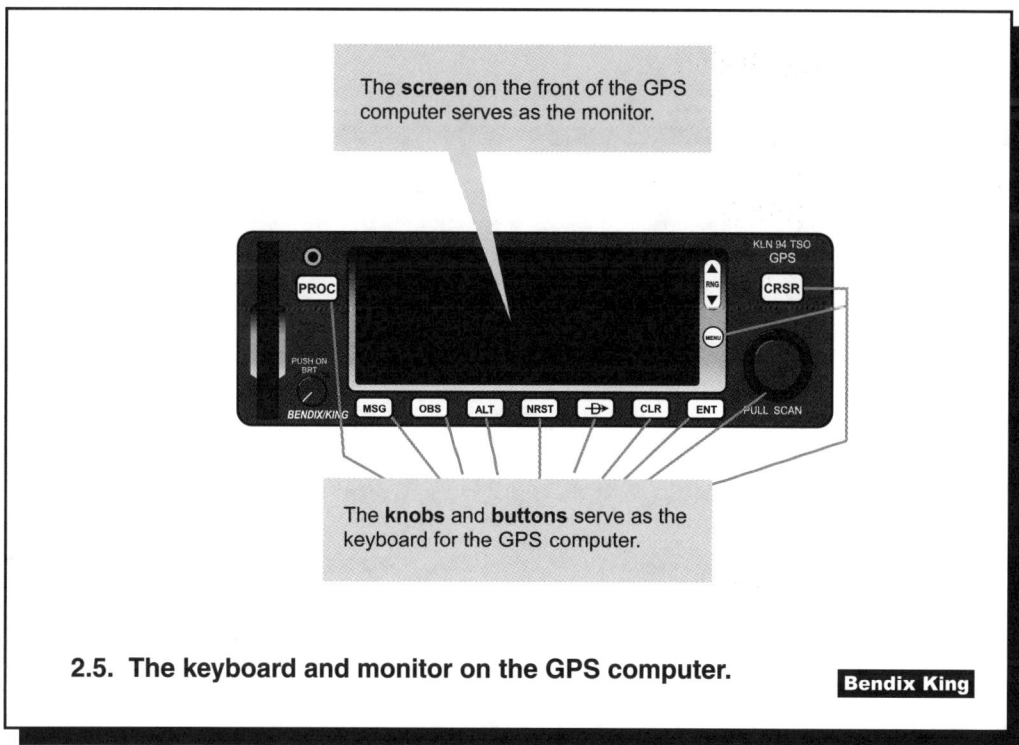

2.5. The keyboard and monitor on the GPS computer.  Bendix King

## Using the Monitor

GPS computers store much more information than can be presented on their small displays. GPS computers organize related information into a collection of separate displays called **pages.** Each page presents information about a particular topic and bears a page title that makes clear the contents of the page. For example, the pages shown in Figure 2-6 present detailed information about airports.

Only one page can be displayed on the monitor at any one time. Figure 2-7 shows how different pages can be displayed on each GPS computer. Note that each computer has its own way of

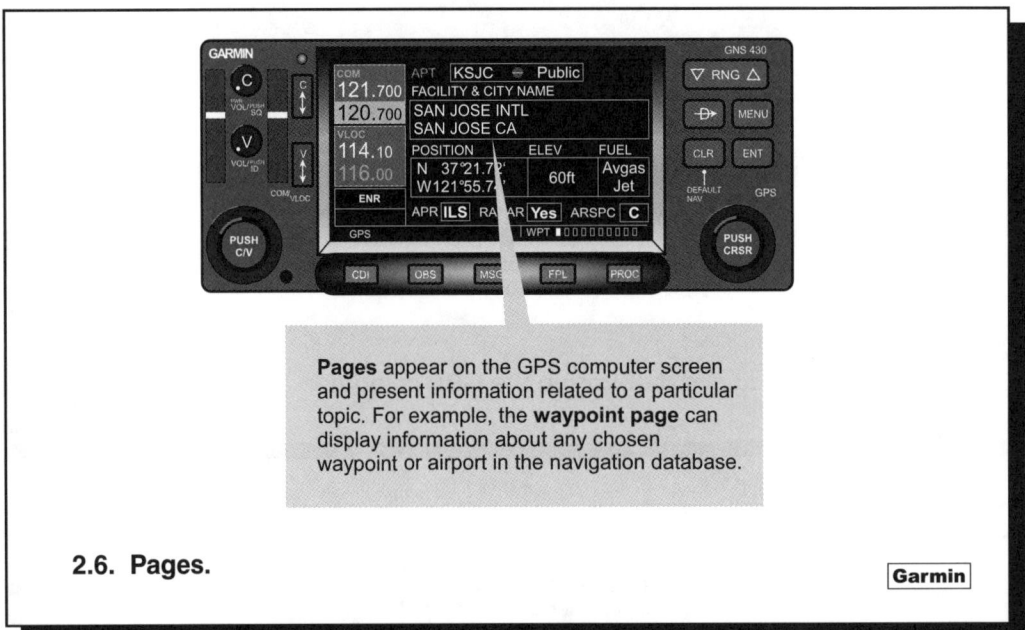

**Pages** appear on the GPS computer screen and present information related to a particular topic. For example, the **waypoint page** can display information about any chosen waypoint or airport in the navigation database.

**2.6. Pages.**  Garmin

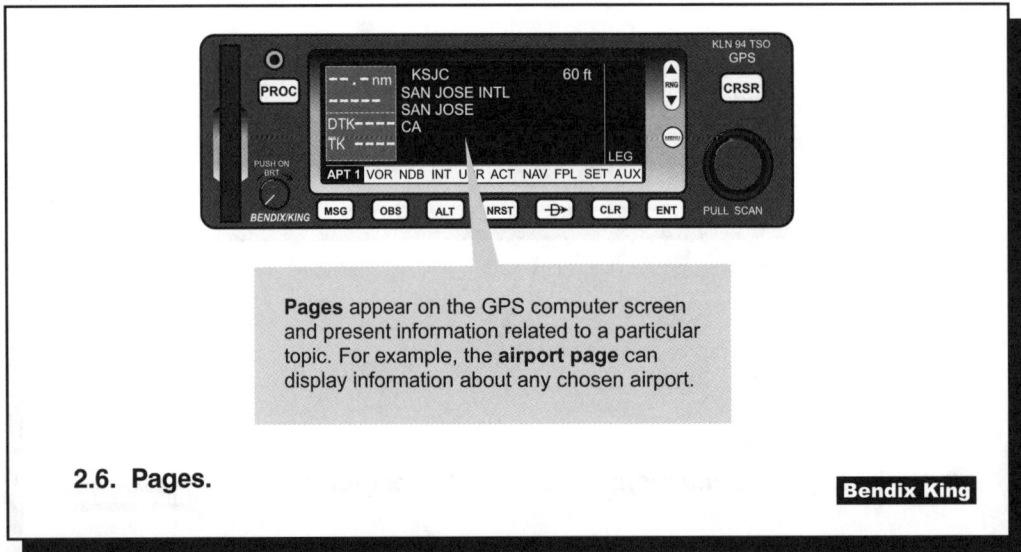

**Pages** appear on the GPS computer screen and present information related to a particular topic. For example, the **airport page** can display information about any chosen airport.

**2.6. Pages.**  Bendix King

## 2 / Planning the Flight Route

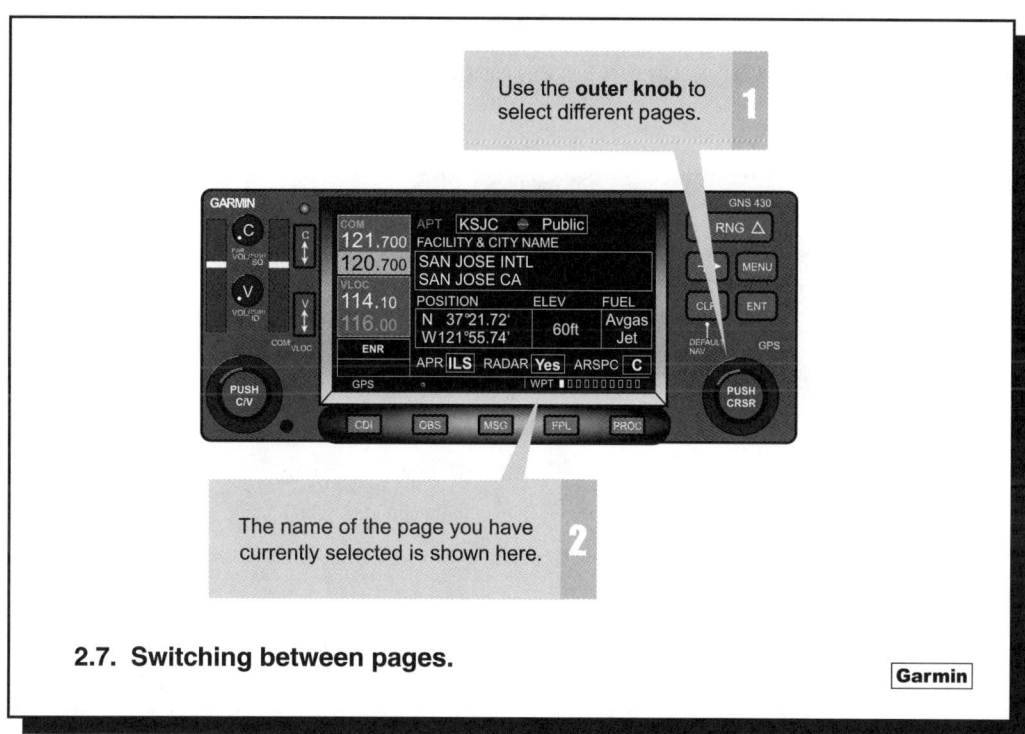

**2.7. Switching between pages.**

reminding you which page you currently have selected. Practice switching between pages and become proficient in the art of page navigation. Being able to find information you need quickly in flight will prove to be a valuable skill.

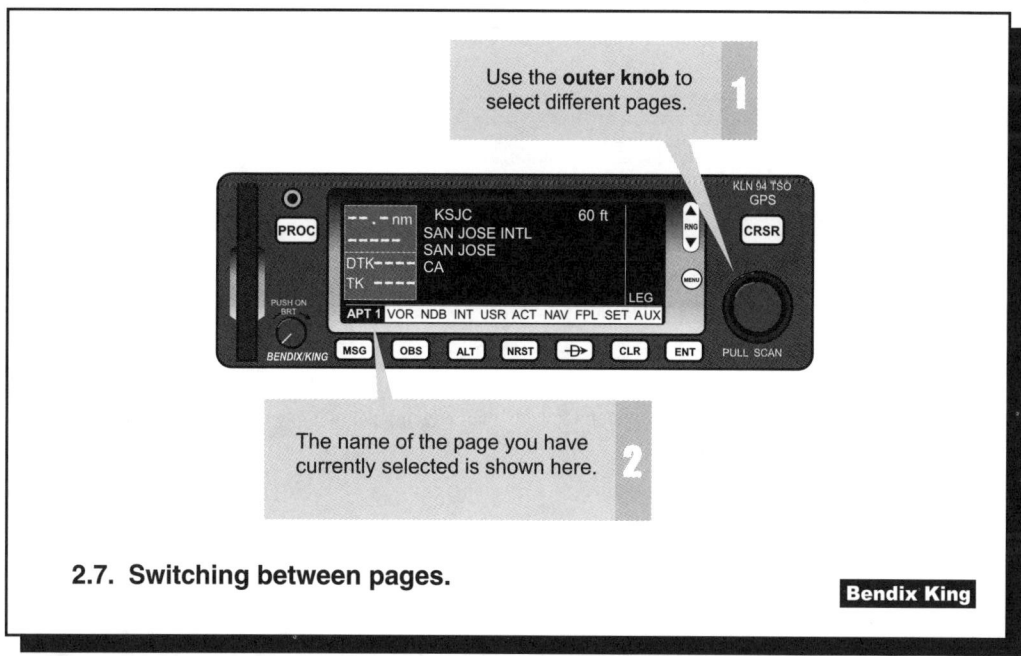

**2.7. Switching between pages.**

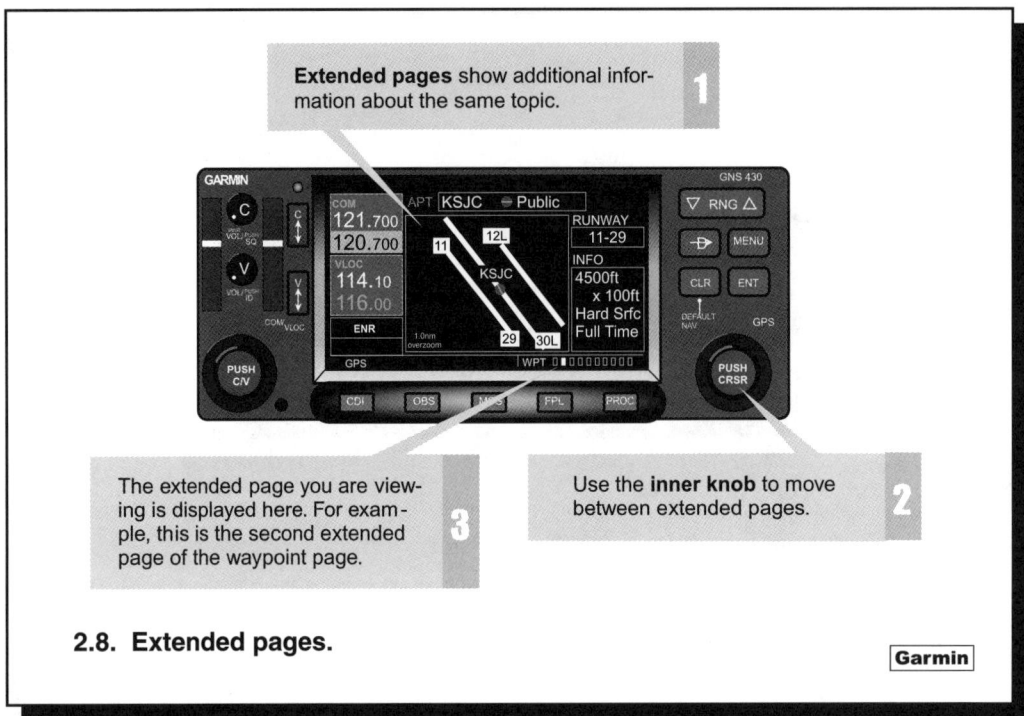

**2.8. Extended pages.**

Most pages contain more information than can fit on the GPS computer monitor. For example, the GPS computer contains a vast amount of information about San Jose International Airport: much more than what appears on the pages shown in Figure 2-7. **Extended pages** solve this problem. Extended pages are essentially pages within pages. Figure 2-8 shows how you can access

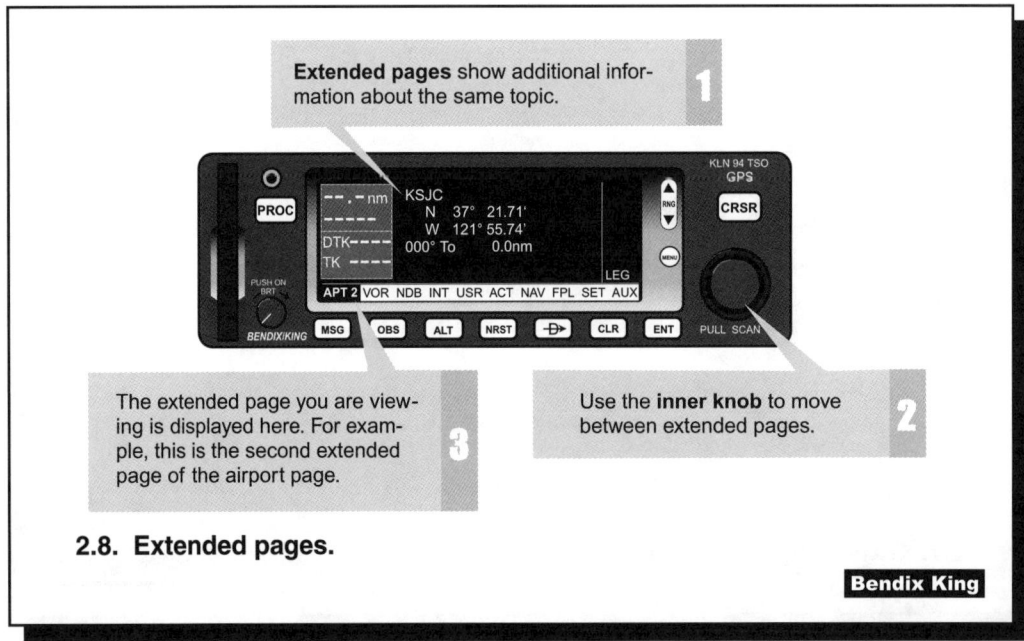

**2.8. Extended pages.**

## 2 / Planning the Flight Route

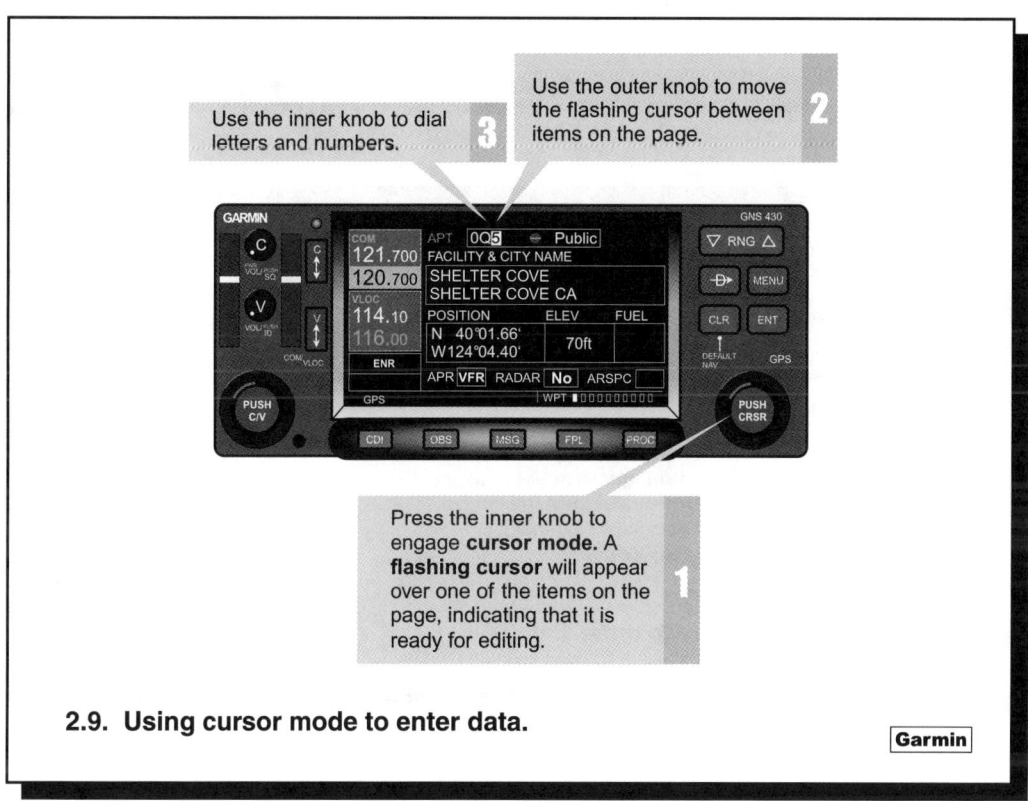

2.9. Using cursor mode to enter data.

extended pages on each GPS computer. Practice switching between the extended pages that contain information about San Jose International Airport. Each of these pages shows different information about the airport such as the latitude and longitude coordinates for the airport, the time zone in which the airport is located, the communication frequencies, and runway names and lengths.

### Using the Keyboard

The buttons and knobs of every GPS computer can be used to enter data into some of the pages that can appear on the computer. Making entries into pages is the way that you input information into the GPS computer when you build and modify your flight plans.

You can make entries on a page using the same knobs you used to switch pages. In order to do this, you first have to place the computer in **cursor mode.** Putting the computer in cursor mode tells the computer that you are now intending to input information. Whenever you wish to input information, you should immediately think of cursor mode. Figure 2-9 illustrates the process of making an entry on a page using cursor mode.

Note that cursor mode changes the way the knobs work on the GPS computer. When in cursor mode, rotating the outer knob no longer moves you between pages. As you rotate the outer knob, some of the items on the page become highlighted and ready for editing. Put the unit in cursor mode and practice rotating the outer knob and watching the cursor move between the items appearing on the page.

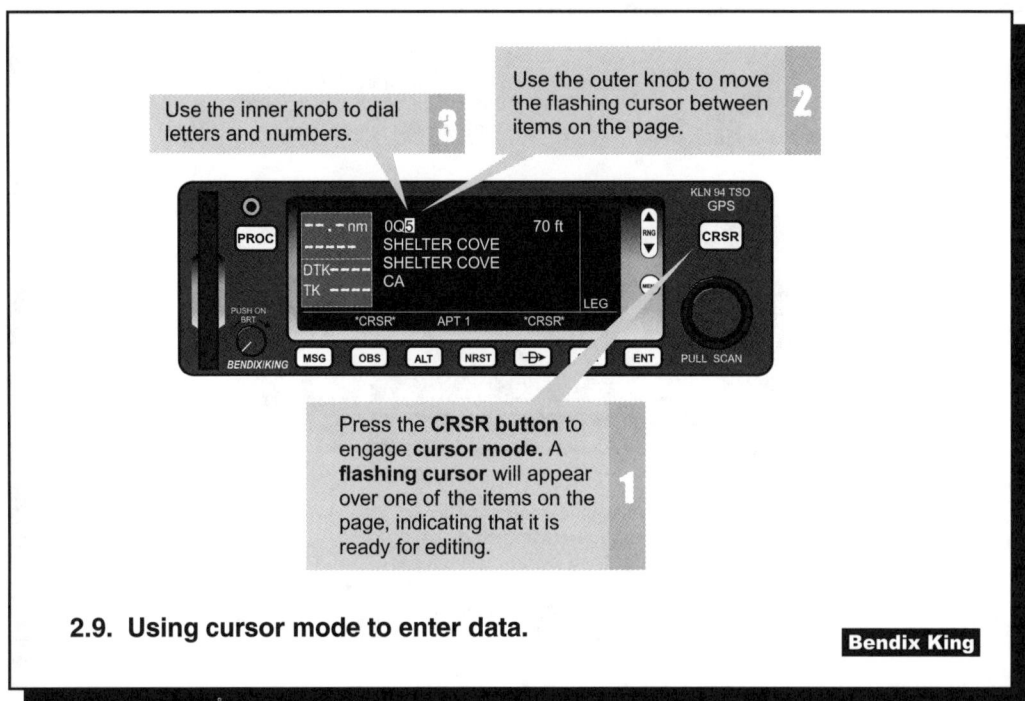

2.9. Using cursor mode to enter data.

Rotating the inner knob while in cursor mode allows you to modify the item that is currently highlighted. You can use the inner knob to scroll among the 26 letters of the alphabet and 10 digits.

Practice using the outer and inner knobs in cursor mode to enter a new airport identifier on the page. Use the outer knob to position the cursor over each character and use the inner knob to change the character. After dialing in each character, rotate the outer knob one position to the right to arrive at the next character field.

Practice entering your favorite airports on the page shown in Figure 2-9. When you are finished entering the characters, get out of cursor mode. Once you are out of cursor mode, the outer knob once again takes you between pages, and the inner knob allows you to scroll between extended pages. To develop your skills, give yourself tasks such as "find the runway length at Shelter Cove (0Q5)."

You may have noticed that this is a rather tedious way to enter data, but, unfortunately, most GPS computers do not have keypads due to size and weight considerations. The confusing aspect of entering data this way is that the inner and outer knobs have different functions depending on the context in which they are being used. Keep working with the knobs and buttons and eventually your fingers will learn their way around.

### Entering the Flight Route

Now that you have learned the basic techniques for getting information in and out of the GPS computer, let's turn our attention to the problem of entering your flight plan into the computer.

Suppose you have planned a flight from San Jose International Airport (KSJC) to Oakdale Airport (O27). You have selected the Sunol Five departure procedure and the GPS Runway 10 approach into Oakdale and have prepared the flight plan shown in Figure 2-1.

## 2 / Planning the Flight Route 21

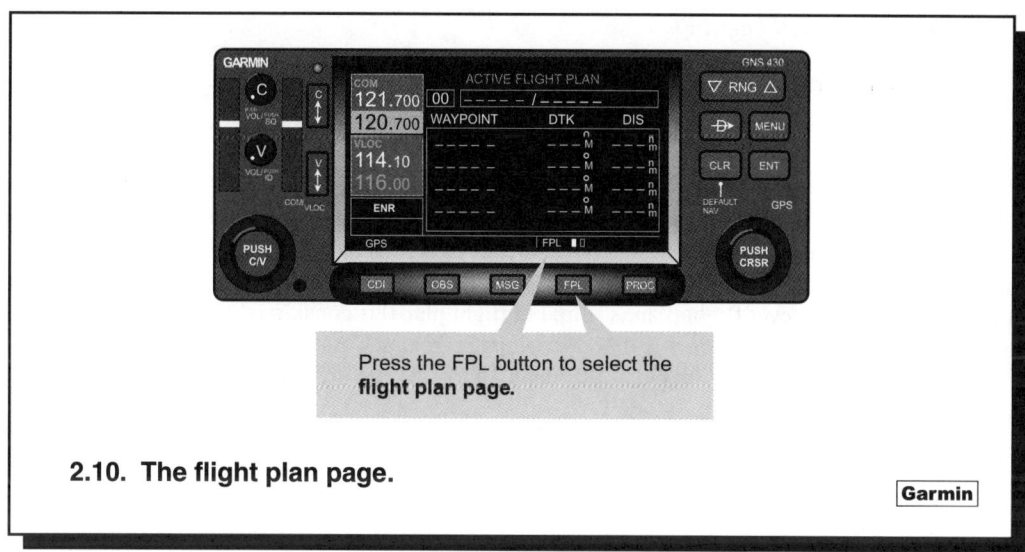

**2.10. The flight plan page.**

You have the nine fixes along your route nicely laid out on your paper flight plan, with all the distances, bearings, times, and fuel burns calculated between them. Your next step is to tell the GPS computer about these choices.

### The Flight Plan Page

Every GPS computer has a page dedicated to the problem of entering a flight plan. Figure 2-10 shows how to access the **flight plan page** using each GPS computer.

Entering a flight route into the GPS computer requires you to fill in the blank lines on this page with the list of waypoints that define your route. There are two ways to accomplish this.

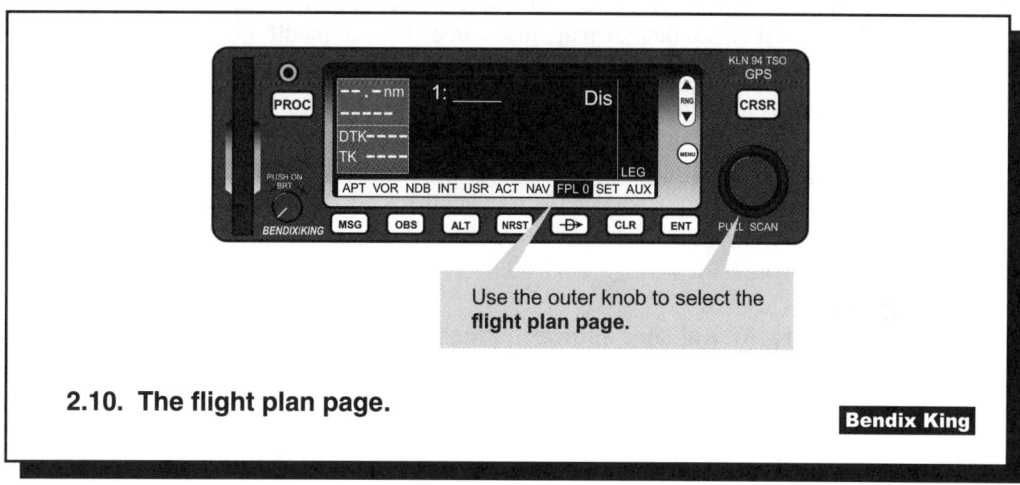

**2.10. The flight plan page.**

### Activating a Stored Flight Plan

The first method of entering a flight route is to use a flight route that has already been built and stored in the GPS computer. Every GPS computer offers a way of preprogramming and storing flight plans for future use. This feature allows you to quickly make use of flight plans that you have taken the time to build in the past.

The two GPS computers discussed in this book each contain a number of memory locations in which flight plans can be stored. These storage locations are named "flight plan 0" through "flight plan 19" (or 25, depending on which computer you are using). Each of these locations can store a single flight plan. Whatever flight plan is stored in flight plan 0 is considered to be the **active flight plan.** The active flight plan is the one that the computer assumes that you wish to navigate to now. The other memory locations, flight plans 1 through 19 (or 25), are used to store flight plans for later use.

In order to use one of the stored flight plans it must first be **activated.** Activating a flight plan means copying the waypoints of a stored flight plan into the flight plan 0 location. Once the stored plan has been copied into flight plan 0, the computer will consider it to be the active flight plan.

Using each of our two GPS computers, Figure 2-11 illustrates how to browse the list of stored flight plans and activate any flight plan of your choosing.

### Entering a Flight Plan by Hand

The second method of building a flight plan requires you to enter the waypoints in your route one by one. You already know how to use the outer and inner knobs to enter characters, so you really only need to dial in the waypoints shown on your paper flight plan in Figure 2-1. But before you do that you need to learn one more thing.

All flight computers such as GPS computers distinguish between two kinds of waypoints: (1) waypoints that are part of a published departure, arrival, or approach procedure, and (2) airports or en route waypoints. GPS computers allow you to manually type in waypoints that are not published procedure waypoints but do not let you type in procedure waypoints. One reason for this is that the consequences of making an error typing in an approach waypoint can be quite serious. Imagine misspelling the name of the final approach fix. Being close to the ground at some mistaken location is never a good idea. Another reason for disallowing you to enter published procedure waypoints is to discourage pilots from making up their own approach procedures or modifying existing ones.

### Entering the Airport and en Route Waypoints

Look at your flight plan and decide which waypoints are not published procedure waypoints and enter them. For your flight, you can enter KSJC, but that's all. All the rest of the waypoints are part of the Sunol Five departure or Oakdale GPS approach. Figure 2-12 illustrates the process of entering waypoints directly into the active flight plan location.

## Entering the Departure, Arrival, and Approach Procedure Waypoints

So how do you enter waypoints that are part of a published procedure? Every IFR-capable GPS computer offers the pilot a menu of published procedures to choose from. Choosing one of these procedures causes the computer to automatically insert all of the waypoints included in that procedure into the flight plan page.

2 / Planning the Flight Route

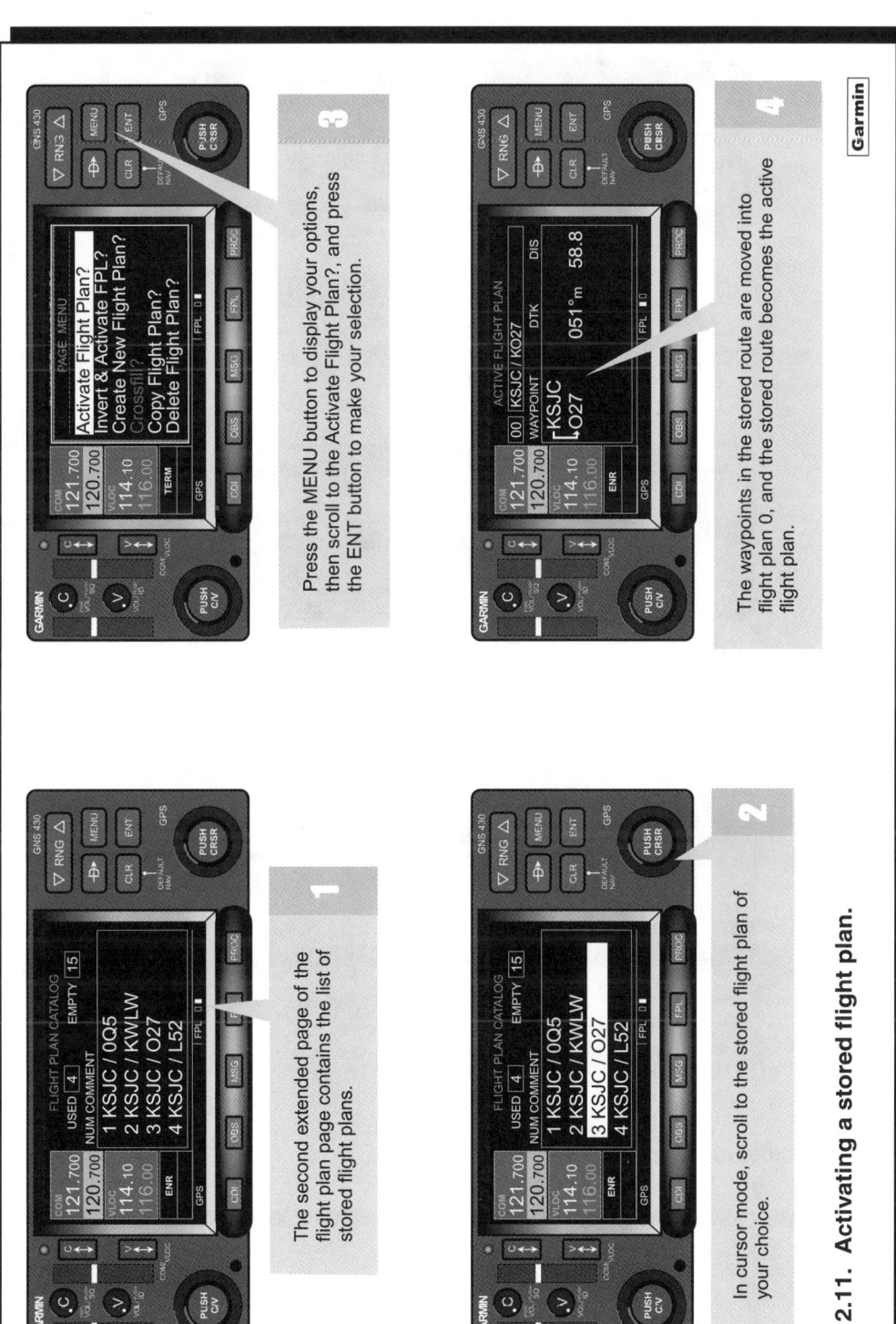

2.11. Activating a stored flight plan.

# 24  COCKPIT AUTOMATION for General Aviators and Future Airline Pilots

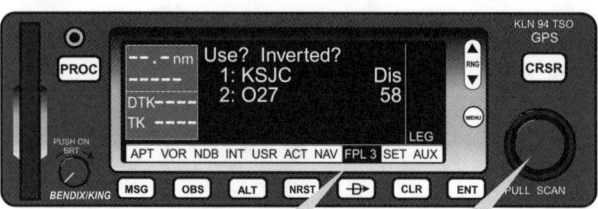

Flight plans can be stored in FPL 1 through FPL 24 (the extended pages of the flight plan page). Use the inner knob to move between the stored flight plans.

**1**

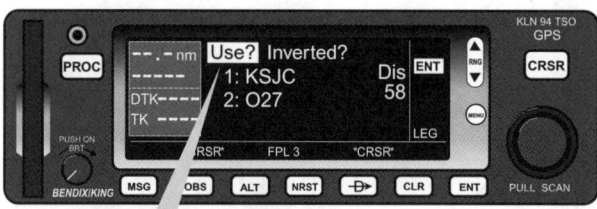

Switching to cursor mode, highlight the Use? option, and press the ENT button.

**2**

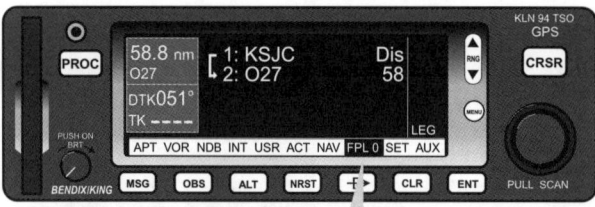

The waypoints in the stored route are inserted into flight plan 0, and the stored route becomes the active flight plan.

**3**

**2.11. Activating a stored flight plan.**

Bendix King

## 2 / Planning the Flight Route

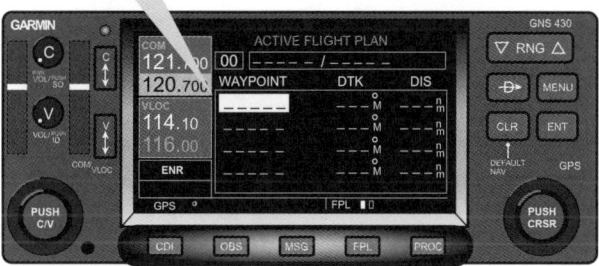

2.12. Building a flight plan from scratch.

Garmin

**26** COCKPIT AUTOMATION for General Aviators and Future Airline Pilots

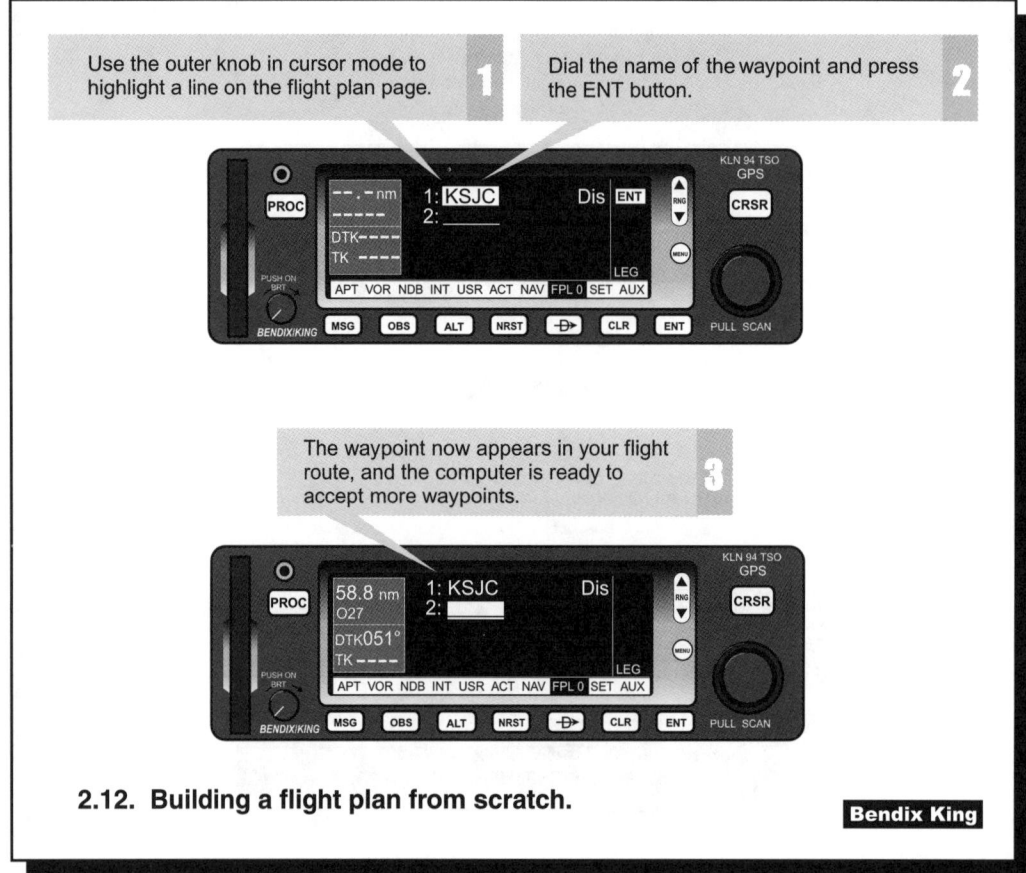

2.12. Building a flight plan from scratch.

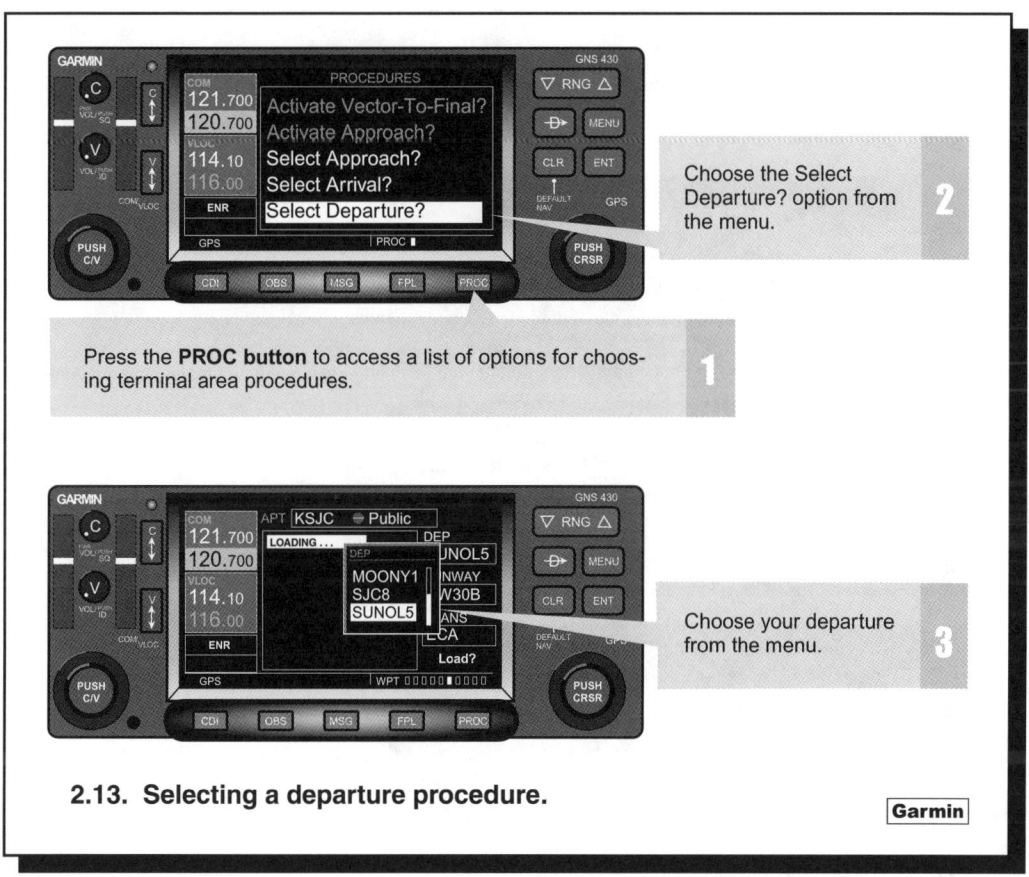

2.13. Selecting a departure procedure.

Figure 2-13 shows you how to bring up the menu of departure procedures for your departure airport. Loading a departure procedure for an airport amounts to nothing more than getting into cursor mode and selecting the desired procedure from the menu.

After selecting the departure, the GPS computer then offers you a menu of transitions and runways for that procedure, as shown in Figure 2-14.

After making your selections, you are then asked if you would like to add the Sunol Five procedure to your flight plan. Figure 2-15 illustrates adding the procedure to your flight plan.

You can include the Oakdale GPS Runway 10 approach in your flight plan in a similar way. Figure 2-16 illustrates the process of loading an approach, which is identical to the process of loading a departure procedure.

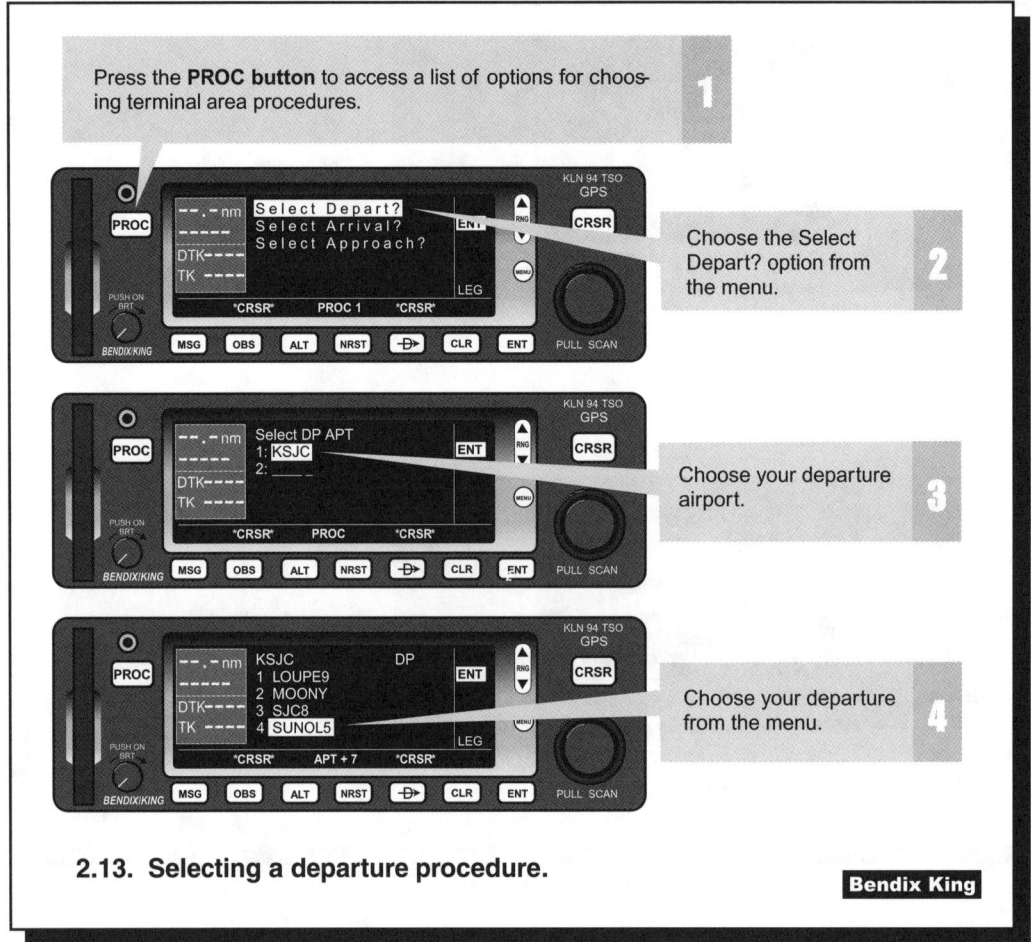

2.13. Selecting a departure procedure.

## 2 / Planning the Flight Route

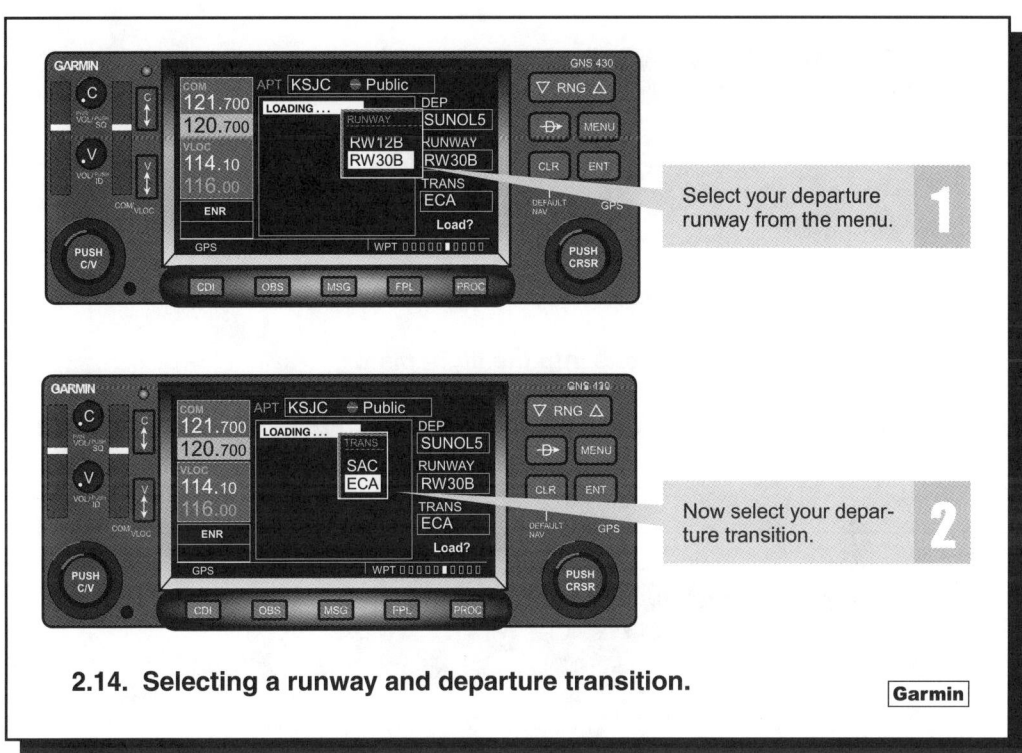

2.14. Selecting a runway and departure transition.

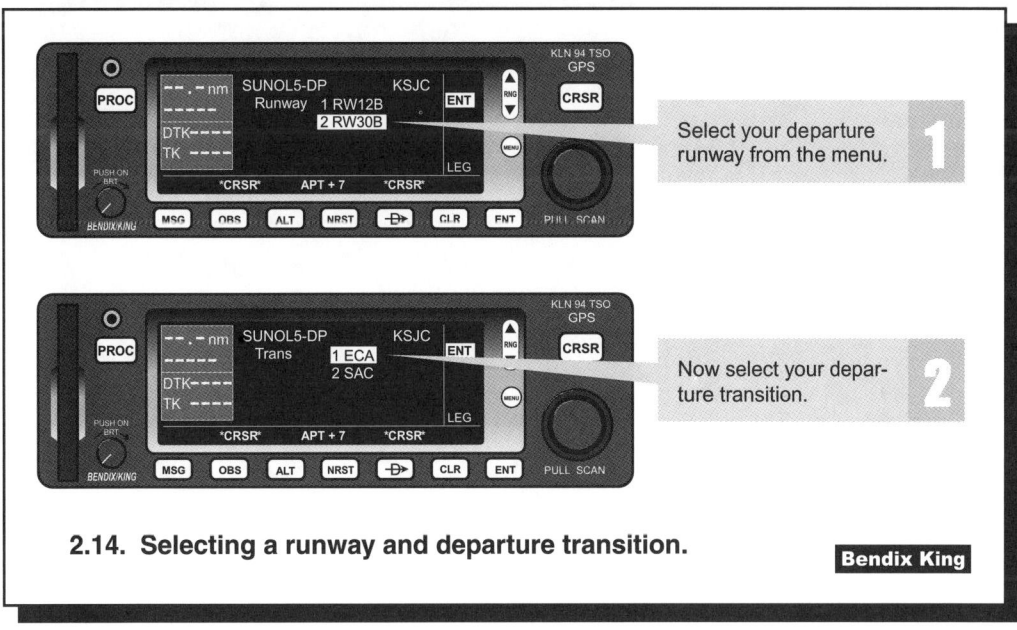

2.14. Selecting a runway and departure transition.

# COCKPIT AUTOMATION for General Aviators and Future Airline Pilots

2.15. Loading your selections into the flight plan.

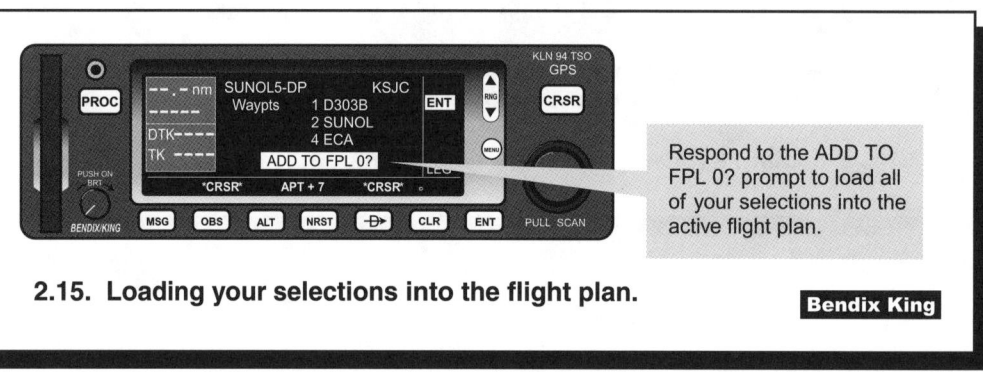

2.15. Loading your selections into the flight plan.

# 2 / Planning the Flight Route

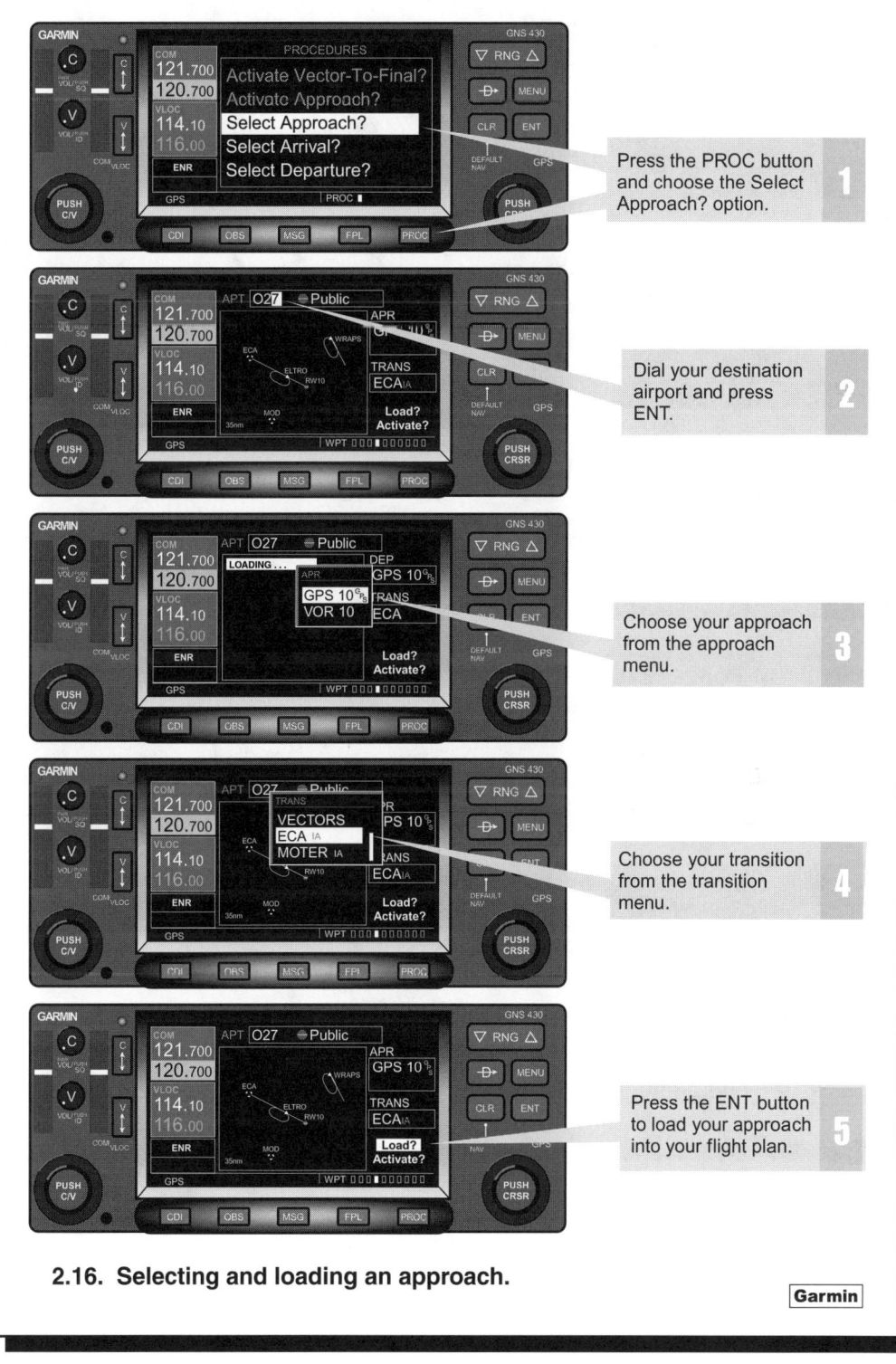

2.16. Selecting and loading an approach.

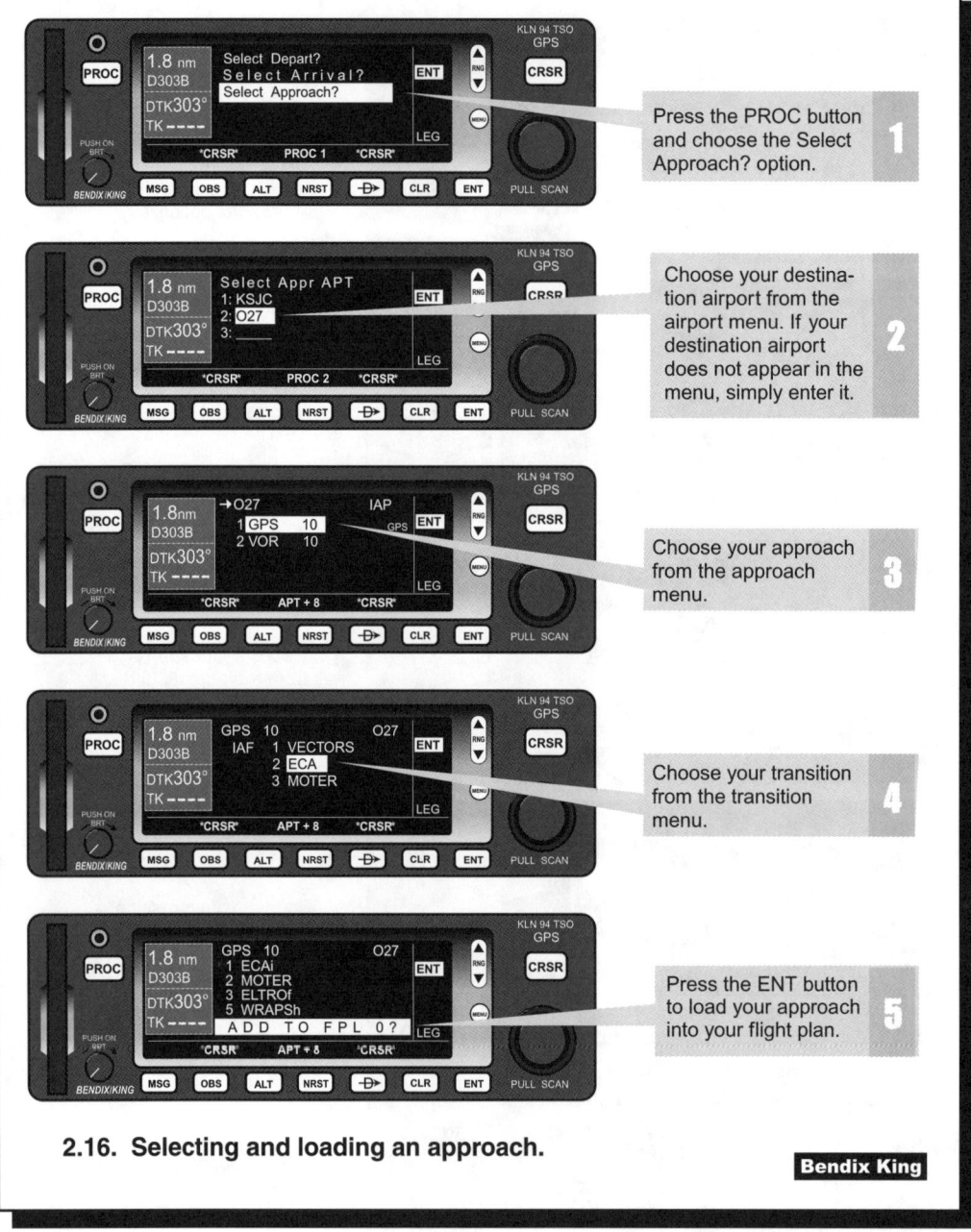

2.16. Selecting and loading an approach.

2 / Planning the Flight Route                                                                   33

## SKILL 3: CHECK THE FLIGHT PLAN TO ENSURE THAT YOU AND THE COMPUTER HAVE THE SAME PLAN

Your next step is to review the route that has been built to ensure it is indeed what you want. It is particularly important to ensure that the programmed route agrees with the clearance you have been issued and with your en route and terminal area charts. You need to check all of the same things you checked when you planned your route using the traditional method. Just because you used a computer to help you build your route doesn't mean that no mistakes or oversights were made.

You can examine the flight route that has been built by switching back to the flight plan page. It's a good idea to check at least three things on the flight plan page when reviewing your route.

### Check the Waypoints

Figure 2-17 shows how to scroll through the waypoints in your route. Compare the sequence of waypoints with the one on your paper flight plan. Are any waypoints missing? Did you mistakenly include any extra waypoints in the route? Did you misspell any waypoints? Did the computer mistakenly include any extra waypoints in the route?

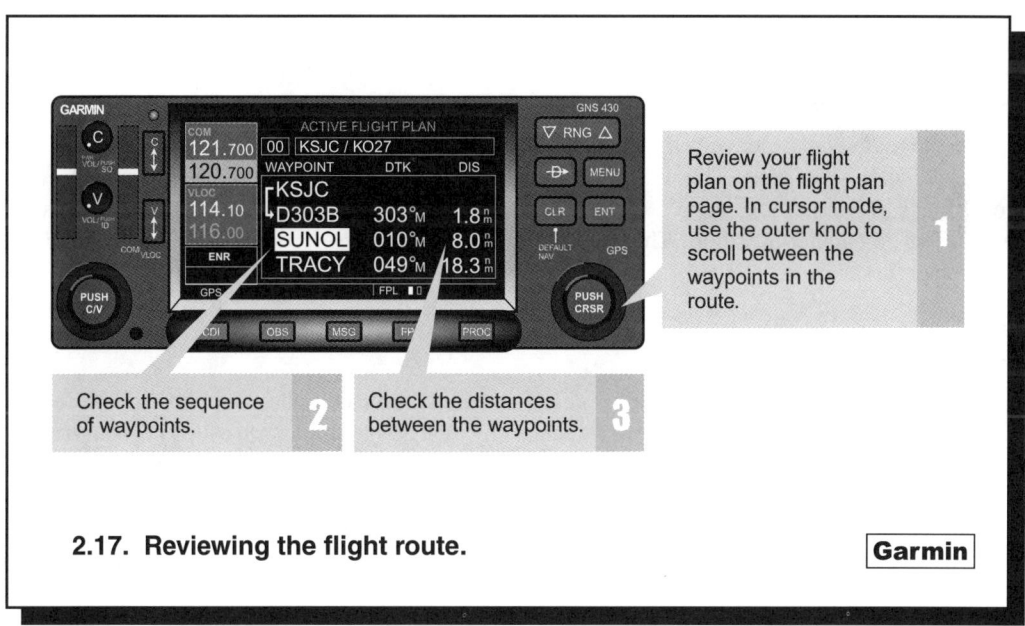

2.17. Reviewing the flight route.

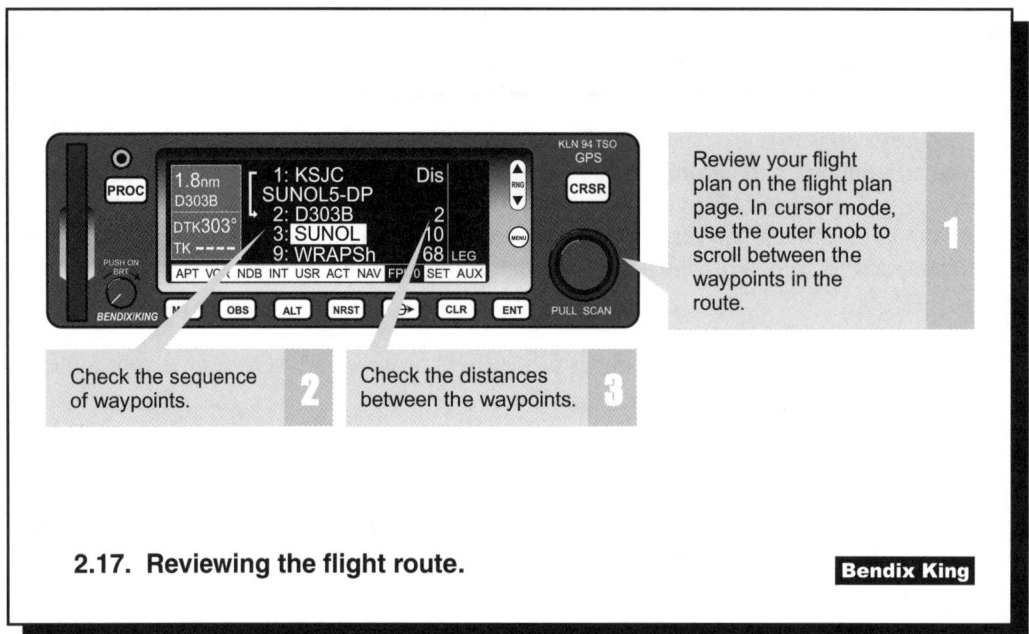

2.17. Reviewing the flight route.

Bendix King

## Check the Distances

On the flight plan page, you can see that the computer has also calculated the distances between the waypoints in your route. These distances can be checked against your en route charts. A common error is to misspell the name of a waypoint and mistakenly enter another waypoint that is located far off in another direction. Checking the waypoint distances for unusual numbers is a good way to spot these errors.

## Check the Desired Tracks

On the flight plan page, you can also see the courses that the computer has chosen between the waypoints along your route. The courses calculated by the GPS computer are called **desired tracks.** Figure 2-18 shows you how to display the desired tracks between the waypoints in your route.

It is interesting to compare the desired tracks with the magnetic courses listed on the published airways that run between the waypoints. You may notice that some of the desired tracks differ slightly from the magnetic courses that appear on your en route charts. There are two reasons for this.

The first reason why the desired tracks in your GPS computer may differ slightly from the courses shown on your charts is that, unlike the world that is printed on paper charts, the earth is round, not flat. The shortest distance between two points on the earth is not a straight line: it is an arc, as shown in Figure 2-19.

2 / Planning the Flight Route 35

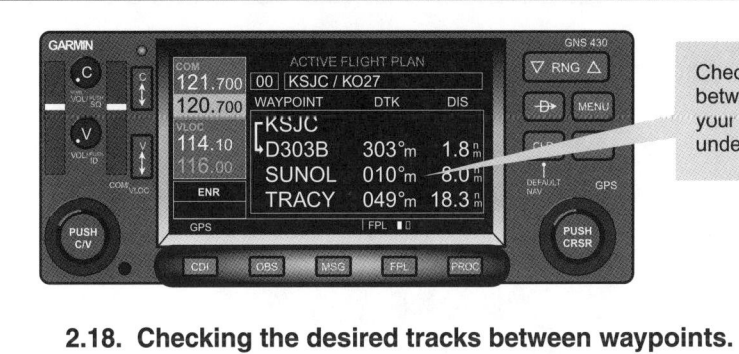

2.18. Checking the desired tracks between waypoints.

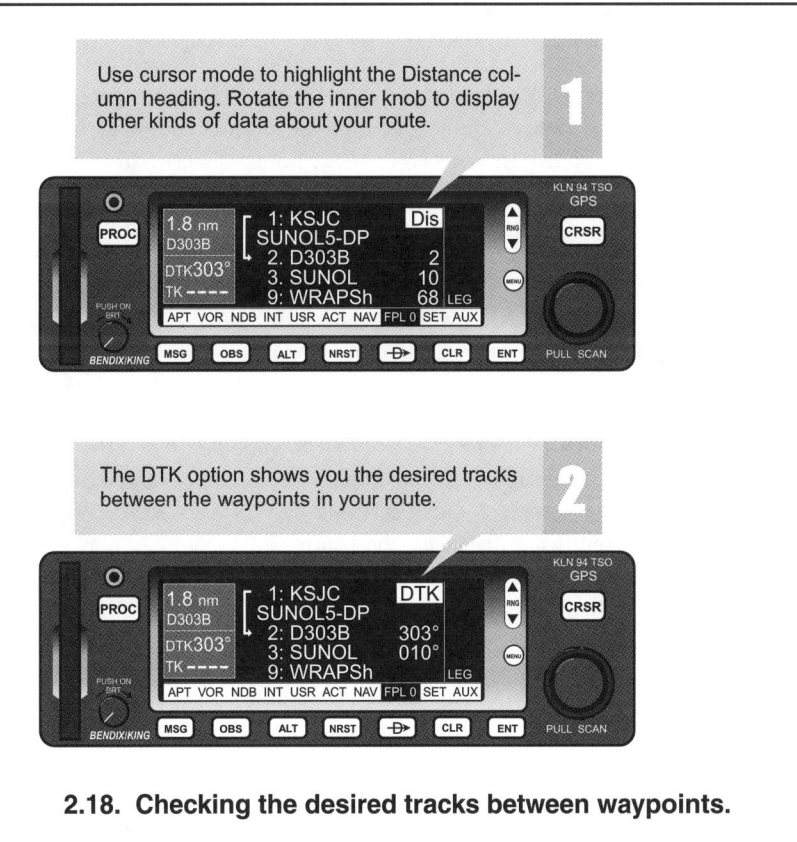

2.18. Checking the desired tracks between waypoints.

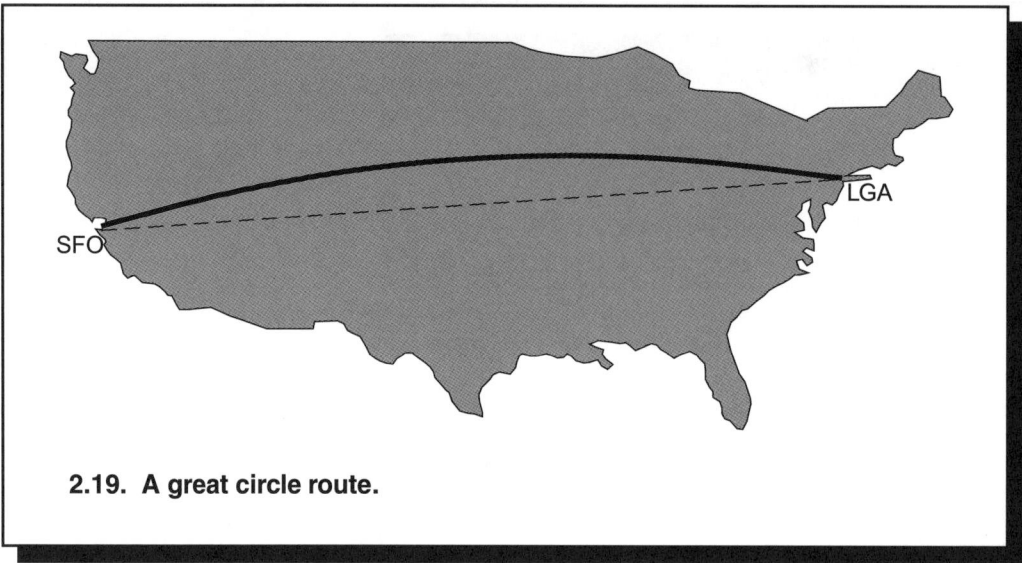

2.19. A great circle route.

The shortest route between two points on the surface of a sphere is called a **great circle route.** Determining a great circle route is rather simple. Just imagine a plane cutting through the earth that passes through three points: your point of origin, your point of destination, and the center of the earth. If your proposed route doesn't lie on this plane, you are taking the long way there.

So how does all this great circle talk explain why your tracks don't match up? This is easy. Look at the route from SFO to LGA in Figure 2-19. What is the bearing on course as you depart SFO? It looks like it is a little less than 90 degrees. How about the bearing as you arrive at LGA? It looks like it is a little more than 90 degrees. The interesting thing to learn from this exercise is that there is no one single bearing to a destination along a great circle route. The bearing is constantly changing, since it's a circle, not a line.

The second reason why the desired tracks calculated by your GPS computer may differ slightly from those published on your charts is that en route charts simplify the problem of correcting for magnetic variation by applying magnetic variations along airways only at each VOR. When the VORs along an airway are spaced at significant distances, the bearing of the airway may change along the way. Your GPS computer is able to apply magnetic variations at every point along your route, making the courses more accurate than those portrayed on the chart, and thus slightly different.

### Filing the IFR Flight Plan

If your aircraft is equipped with an IFR-certified GPS computer, the equipment suffix for the aircraft is "/G." But remember, the /G indicates that the equipment is capable of being used to navigate along a programmed flight route. The real question to consider is whether or not the pilot is capable of operating the equipment. If you don't think you are ready to fly under IFR using any equipment, stick with what you know until you have a chance to learn more about it.

 **PRACTICE SESSION (SIMULATOR)**

Now that you know how to build and review a flight route using your GPS computer, take a few minutes to practice your newly acquired skills. If you haven't already, download the free simulator for your GPS computer on either of the two manufacturers' Web sites.

www.bendixking.com
www.garmin.com

### SKILL 1: Determine If Your GPS Computer Is Approved for IFR Navigation

1. If you have access to an IFR GPS-equipped airplane, find the Form 337 and the supplement in the Pilot Operating Handbook for the GPS computer.
2. Power up the computer and check to see that the navigation database is current.

### SKILL 2: Program the GPS Computer with a Flight Plan

1. Program the same route from San Jose to Oakdale shown in Figure 2-1.
2. Do the same for a flight or two in your area that you might fly sometime.

### SKILL 3: Check the Flight Plan to Ensure That You and The Computer Have the Same Plan

1. Compare each waypoint appearing in the programmed flight route to the en route and terminal area charts that will be used during the flight. Do the distances and bearings agree?

### How Did You Do? (Common Errors)

At the end of this practice session, review each item listed above and write down how well you did. Experience has shown that a few practice runs should allow your fingers to learn their way around the GPS computer. Practice building a route every once in a while to help you remember these skills.

 **DVD DEMONSTRATION DISC**

Skills 1, 2, and 3 are demonstrated using the Garmin GNS 430 on the accompanying DVD disc.

# CHAPTER 3

# Following the Flight Route

*This chapter teaches you how to use the GPS computer to follow a flight route that you have built using your GPS computer: from your departure airport to a GPS approach down to minimums at your destination airport. Along the way, you will master four new skills:*

> SKILL 4: *Ensure adequate GPS signal reception.*
> SKILL 5: *Demonstrate how to follow and monitor your progress along the route stored in the GPS computer.*
> SKILL 6: *Plan and fly a descent.*
> SKILL 7: *Demonstrate a GPS approach explaining all approach mode transitions.*

In this chapter, you learn to use the same GPS computer to follow the flight route that you programmed in the previous chapter. You will see that the GPS computer provides many useful services to you along your route, simplifying the process of making your way between the waypoints listed in your flight plan. You will learn that the GPS computer offers several pages that help you keep track of your progress. Your GPS computer will assist you in planning your descent into the terminal area and then guide you along a new kind of instrument approach in which you will use your GPS computer as your primary means of navigation.

## SKILL 4: ENSURE ADEQUATE GPS SIGNAL RECEPTION

When you set your flight plan in motion, you will see how your GPS computer is able to track the position and movement of your airplane along your planned route. Your GPS computer accomplishes this tracking feat using signals received from a collection of satellites that are in constant orbit around Earth. These satellites, together with the computers that receive and interpret their signals, make up the global positioning system.

Although the global positioning system is highly reliable, satellite reception is sometimes interrupted. Consequently, when using your GPS computer for navigation, you have to ensure at all times that the computer is operational and receiving adequate GPS signals. What if the computer wasn't receiving GPS signals? What if the signals were corrupted? How does the pilot determine this? Fortunately, all GPS computers that are approved for IFR navigation have an automated feature that continually checks on the status of the GPS signals that the computer is using to determine the position of the aircraft. This function is called **receiver autonomous integrity monitoring** (or **RAIM**). RAIM works self-sufficiently and notifies you when there is a problem with your GPS signal reception. RAIM requires adequate reception of at least five GPS satellites at all times to consider the computer reliable for IFR navigation. When there is a problem with reception, your GPS computer provides you with an alert.

Figure 3-1 shows the RAIM alerts given by each kind of GPS computer in the event of inadequate GPS reception. These messages clearly state that there is a problem with the GPS reception and that the aircraft position information provided by the GPS computer should no longer be considered reli-

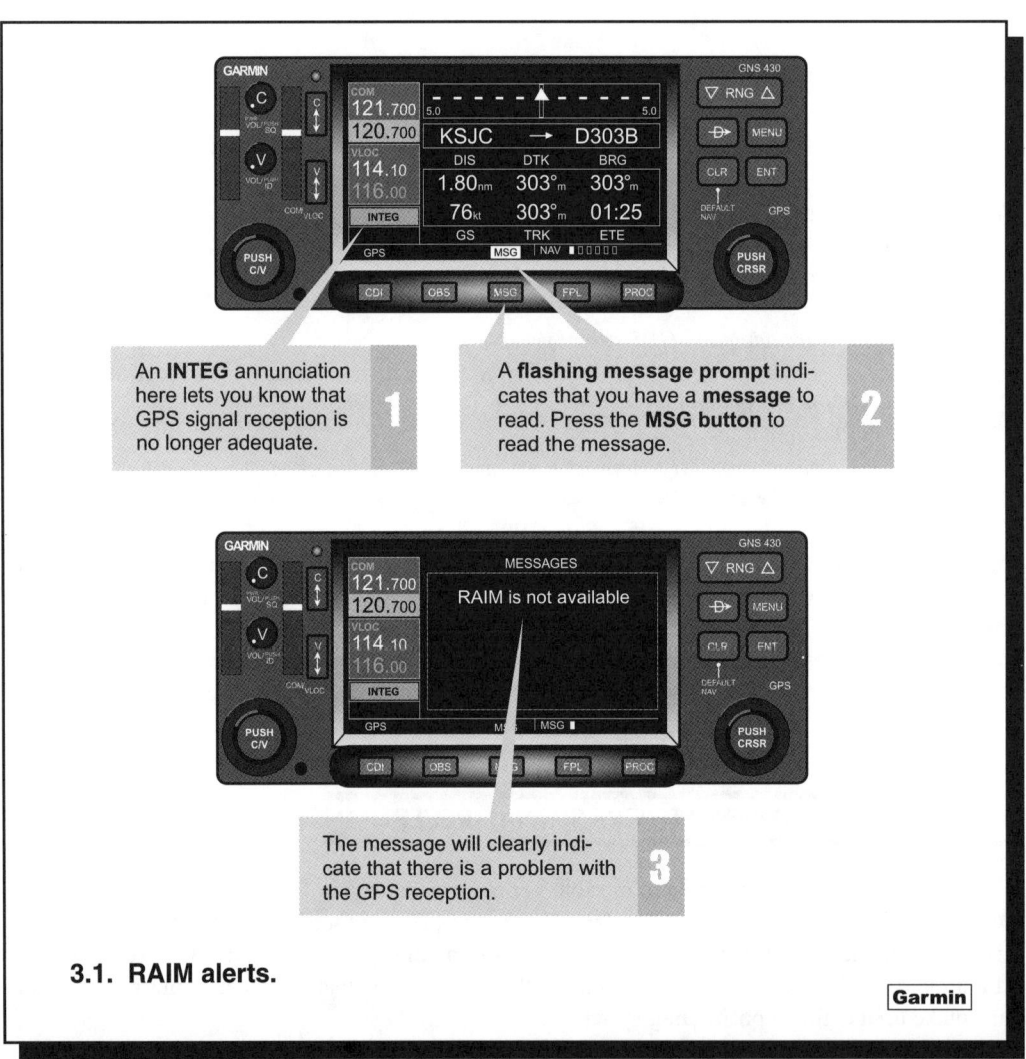

3.1. RAIM alerts.

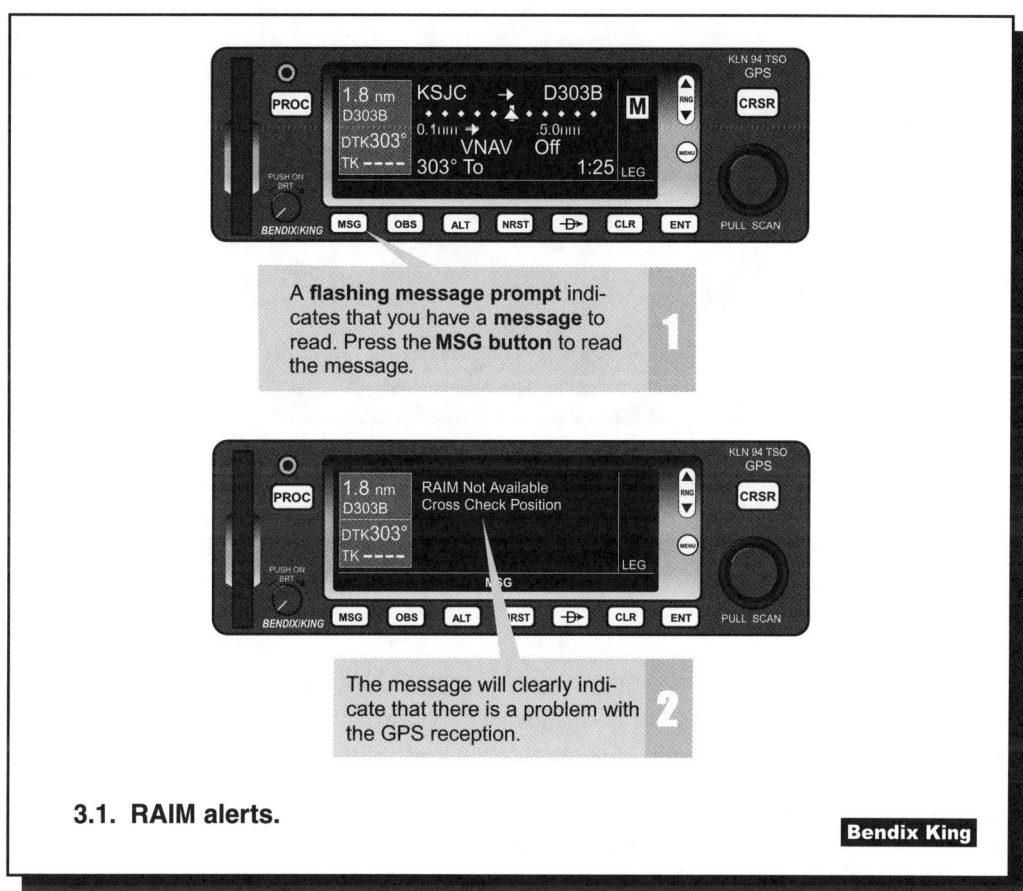

3.1. RAIM alerts.

able. The regulations require airplanes that are equipped with an IFR GPS computer to have alternate means of IFR navigation onboard (VOR receivers are a popular choice). As every pilot knows, you should always make use of every available means of navigation onboard your airplane. When a RAIM message is received, using your alternate means of navigation may be your only way home.

## SKILL 5: DEMONSTRATE HOW TO FOLLOW AND MONITOR YOUR PROGRESS ALONG THE ROUTE STORED IN THE GPS COMPUTER

Looking at the flight plan page in Figure 3-2, you can see that the route stored in the GPS computer is nothing more than a sequence of waypoints. That is, the GPS computer has built a route by playing a game of connect-the-dots between the waypoints that make up the route and terminal procedures you have chosen.

The next part of your job is to follow the flight route that you have planned. As you fly along your planned flight route, the GPS computer will provide guidance from waypoint to waypoint. At any given time, you are making progress to the next waypoint in the programmed flight route. This next waypoint is called the **active waypoint**. In Figure 3-2, the active waypoint is D303B. On the flight plan page, the active waypoint is always the one with the arrow pointing at it.

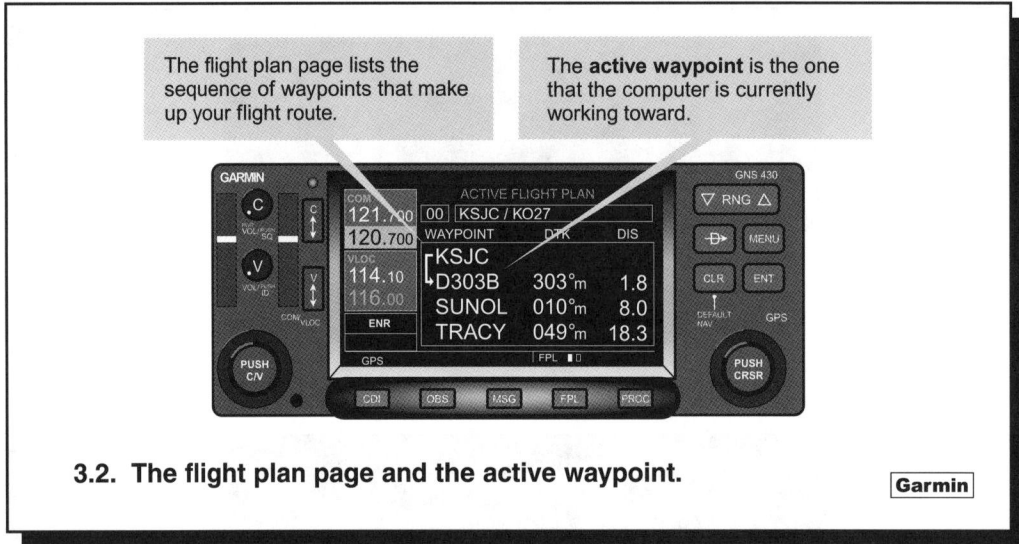

**3.2. The flight plan page and the active waypoint.**

## The Navigation Page

How can the GPS computer help guide you to the active waypoint? Every GPS computer has a page called the **navigation page.** The navigation page, shown in Figure 3-3, keeps you informed about the position of the aircraft and the progress being made toward the active waypoint.

Along with the name of the active waypoint, the navigation page also shows your **desired track** to the active waypoint. The desired track is the track that connects the waypoint you just flew from to the active waypoint. The desired track represents the course that you should be flying on. On the navigation page in Figure 3-3, the current desired track is the 303-degree course between the KSJC and D303B waypoints.

The navigation page also displays a **course deviation indicator** (**CDI**). This familiar-looking display shows your position relative to the desired track to the active waypoint. The CDI on the

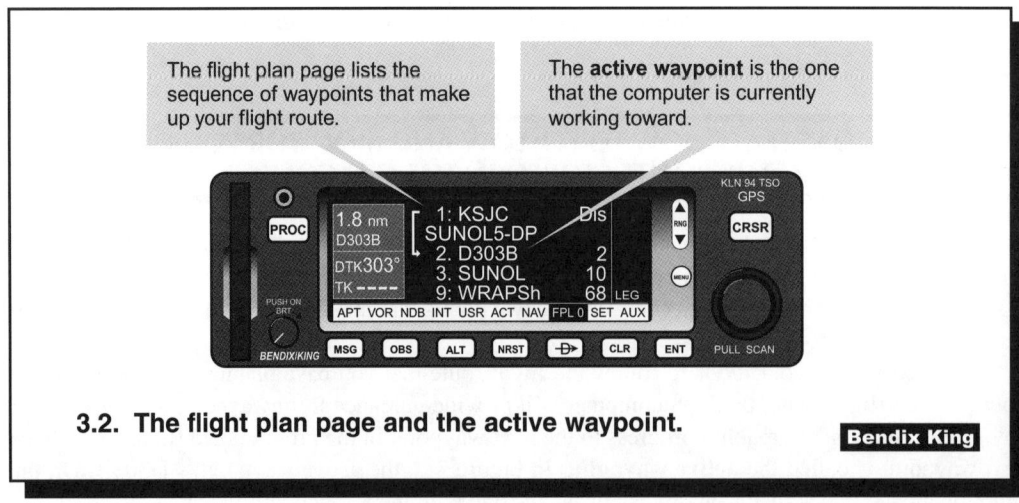

**3.2. The flight plan page and the active waypoint.**

# 3 / Following the Flight Route

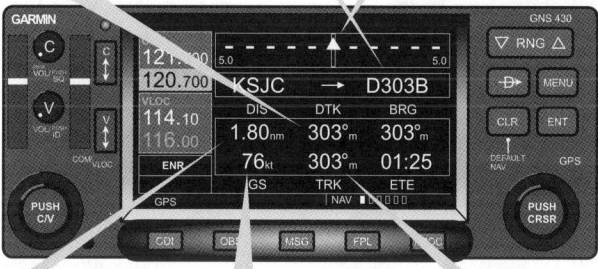

3.3. The navigation page.

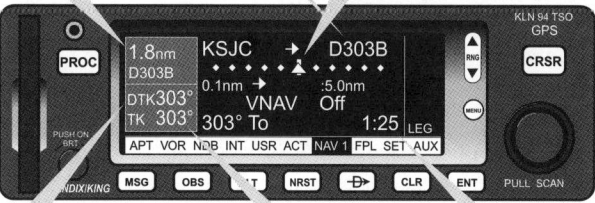

3.3. The navigation page.

navigation display is accompanied by a scale that indicates the distance off course you must fly in order to see a full-scale deflection of the CDI needle.

The navigation page also displays the **track** of the aircraft over the ground. This number tells you in which direction the airplane is really flying. Note that the track is the result of the heading of the aircraft together with the winds that are acting on the aircraft at any given time. Given winds, it is likely that your track and heading are different. Note that you can get a very good sense of what the winds are doing by comparing the track and heading of the aircraft. If you are flying a heading of 090 degrees and the track is 080 degrees, you know that the winds are coming from the south. Notice that having a track indication trivializes the problem of maintaining the airplane on the desired track. Leaving San Jose airport, to follow the 303-degree desired track to D303B, simply fly a track of 303 degrees. With a track display, you no longer have to rely on the guesswork of "bracketing" to figure out which heading will result in a track that matches the desired track. The heading indicator is sometimes referred to as the "cockpit wind sock" since it serves as a handy tool for maintaining wind awareness, while the track display is used to establish the aircraft on the proper course.

The navigation page also shows your **ground speed.** Again, the GPS computer eliminates the need to calculate ground speed using true airspeed and estimated winds. Based on your ground speed and distance from the active waypoint, the navigation page also provides you with an **estimated time of arrival** at the active waypoint.

### Reaching the Active Waypoint

Departing San Jose, you will soon reach the D303B waypoint. What happens then? It seems that you would like to forget about flying toward D303B (since you've already been there and done that) and move on to the next waypoint in the sequence: Sunol. To do that, you'll need to realize when you are about to reach the D303B waypoint, lead your turn a bit to not overshoot the course to Sunol, and then start flying on the desired track to Sunol. It turns out that the GPS computer can help you with all three of these tasks.

Most IFR-capable GPS computers offer two different modes of operation when navigating along a route that you have programmed, only one of which can be used at a time. In this chapter, we will use the simpler of these two navigation modes. You will learn how this so-called **sequencing mode** greatly simplifies the process of making your way between the waypoints in your programmed route. The sequencing mode is called **Leg mode** on the Bendix/King GPS computer and **Sequence mode** on the Garmin computer. Figure 3-4 shows you how to engage the sequencing mode on each GPS computer. Figure 3-4 also shows you a different extended navigation page. This extended page shows you the present position of the aircraft plotted against a graphical depiction of your programmed flight route. Check your manufacturer's manual to read about the many features of the "map display."

The sequencing mode simplifies the problem of steering the aircraft along the programmed route by performing three interesting services for you during your flight.

### Waypoint Alerting

The first service performed by the sequencing mode is called **waypoint alerting.** Just prior to reaching each active waypoint, waypoint alerting advises you of your imminent arrival at the active waypoint. Figure 3-5 shows how each GPS computer provides waypoint alerting as you approach Sunol intersection.

3 / Following the Flight Route 45

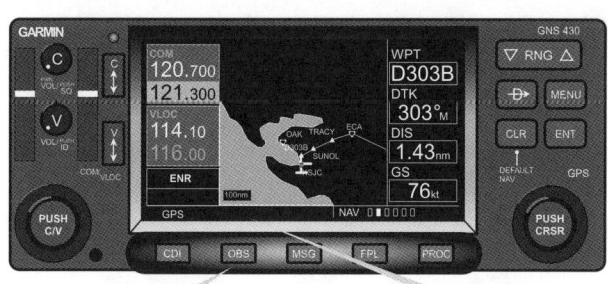

3.4. Engaging the sequencing mode.

Garmin

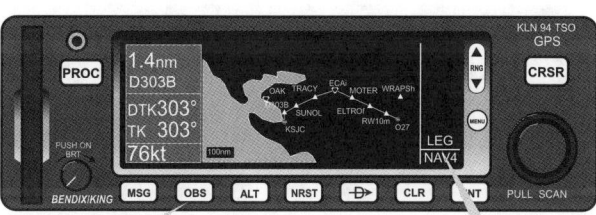

3.4. Engaging the sequencing mode.

Bendix King

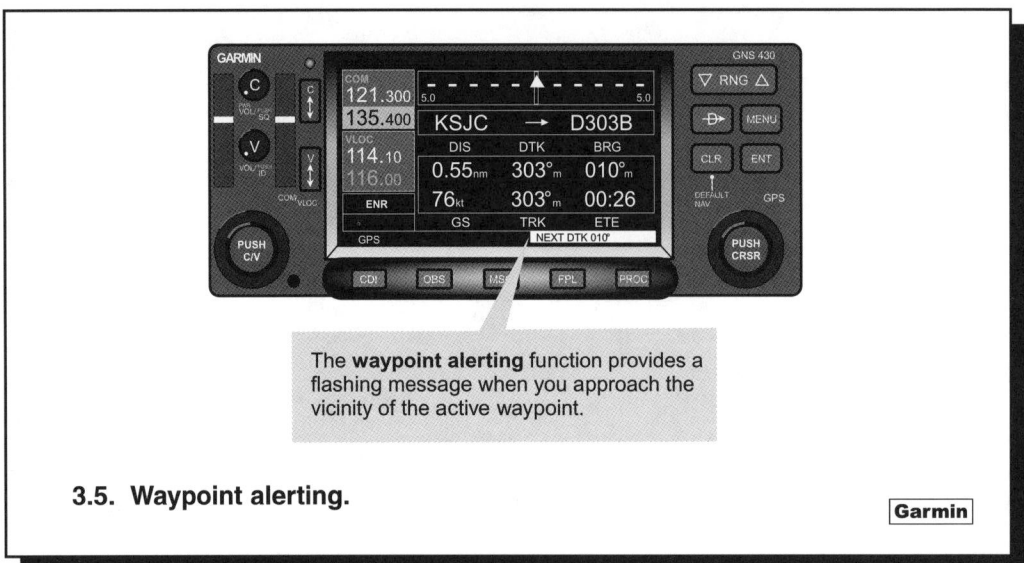

**3.5. Waypoint alerting.**

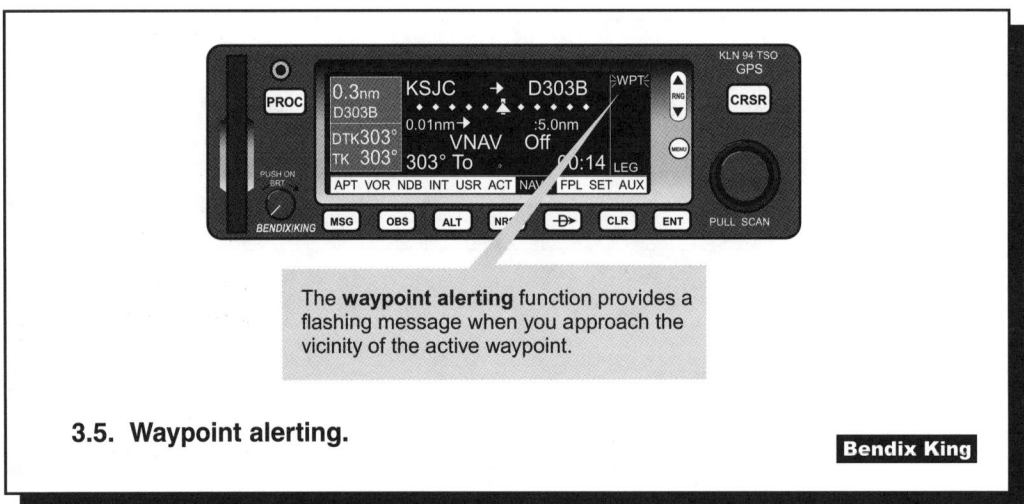

**3.5. Waypoint alerting.**

*Turn Anticipation*

The second service performed by the sequencing mode is called **turn anticipation.** During waypoint alerting and prior to reaching the active waypoint, each GPS computer will provide you with an indication that it is time to begin your turn to fly the desired track to the new active waypoint. The timing of turn anticipation is based on the aircraft's observed ground speed and the angle of the turn required to track to the next waypoint. If you begin a standard rate turn when the way-

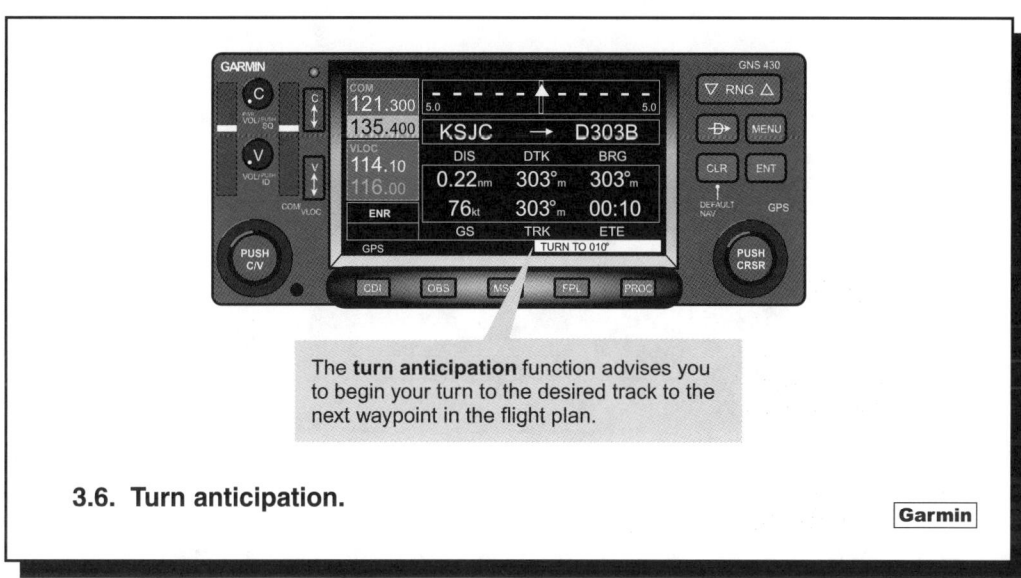

3.6. Turn anticipation.

point alerting indication is presented, you should roll out on course when the aircraft reaches the center of the desired track to the new active waypoint. Figure 3-6 illustrates how each GPS computer provides turn anticipation alerts.

Note that the waypoint alerting and turn anticipation features are intended to be used as backups for alert pilots who remain aware of their position with respect to the flight plan at all times. In twenty years of experience with automation, we have learned that overreliance on features such as waypoint alerting have led to many pilot-automation interaction problems. The use of features such

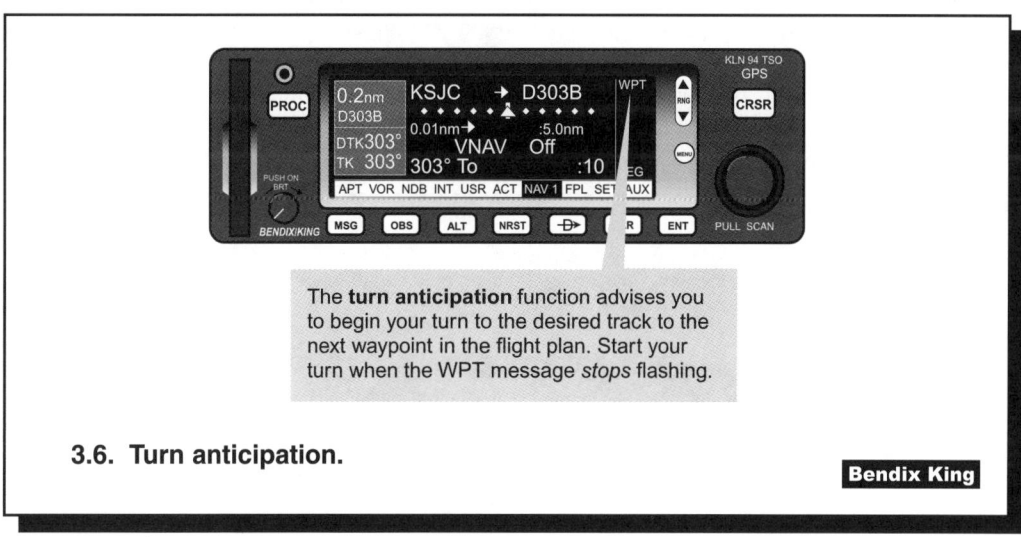

3.6. Turn anticipation.

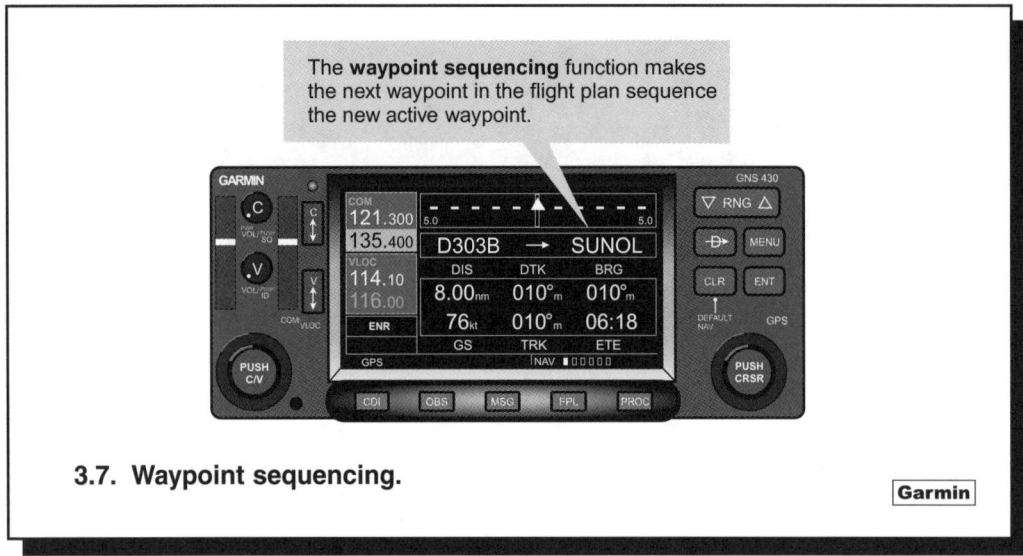

3.7. Waypoint sequencing.

as waypoint alerting can have the effect of deteriorating your position awareness. This problem becomes especially important when the GPS computer has been misprogrammed or is malfunctioning. You wait for an alert that never comes and then stray from the assigned route.

*Waypoint Sequencing*

The third service performed by the sequencing mode is called **waypoint sequencing.** Once you have reached the active waypoint, the GPS route automatically makes the next waypoint in the flight plan sequence the new active waypoint. Waypoint sequencing is illustrated in Figure 3-7.

Turn anticipation and waypoint sequencing are only provided in the sequencing mode. Be advised that turn anticipation and waypoint sequencing services are not provided when the computer is not in the sequencing mode. You can easily determine whether or not the computer is operating in the sequencing mode by checking the mode annunciations illustrated in Figure 3-4.

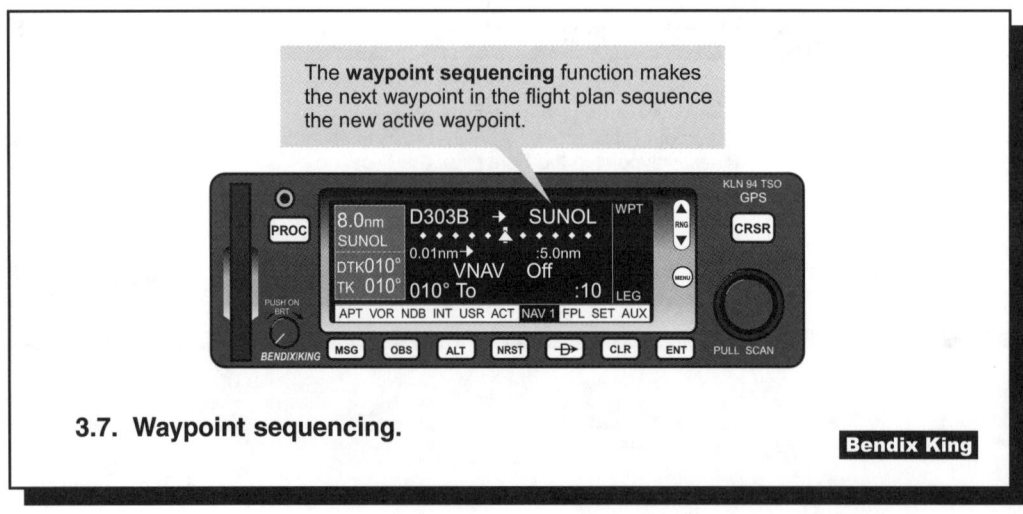

3.7. Waypoint sequencing.

## GPS and the Instrument Scan

As you have just seen, the GPS computer offers much useful information as you make your way along your flight route. This raises the issue of how to include the GPS computer in your normal instrument scan. Remember that while the GPS computer provides you with additional information, it does not relieve you in any way from your existing instrument scan duties. No GPS computer will help you keep your airplane in a desired attitude. Including the GPS computer in your scan is a good idea and one that will take a bit of practice. Be careful to avoid any temptation to stare at the GPS computer or be lured away from scanning your instruments. There is a cost associated with the information provided by the GPS computer: You have to devise a procedure for safely including it in your scan.

## Slaving a Navigation Indicator to the GPS Computer

Most GPS computers allow a **navigation indicator,** conventionally used to show VOR course indications, to be slaved to the GPS computer. If your airplane is so equipped, you can slave the indicator to the GPS computer, as illustrated in Figure 3-8.

Note how each GPS computer has a way of indicating to which receiver the navigation indicator is currently slaved. It is your responsibility to remain aware of what indications are being

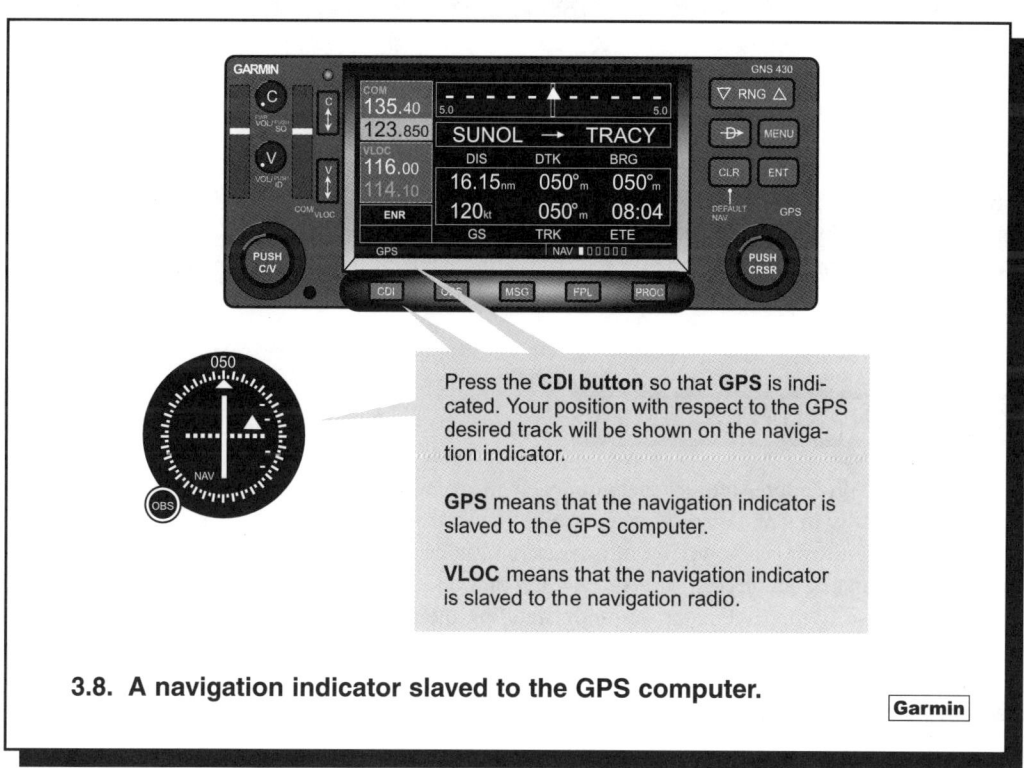

3.8. A navigation indicator slaved to the GPS computer.

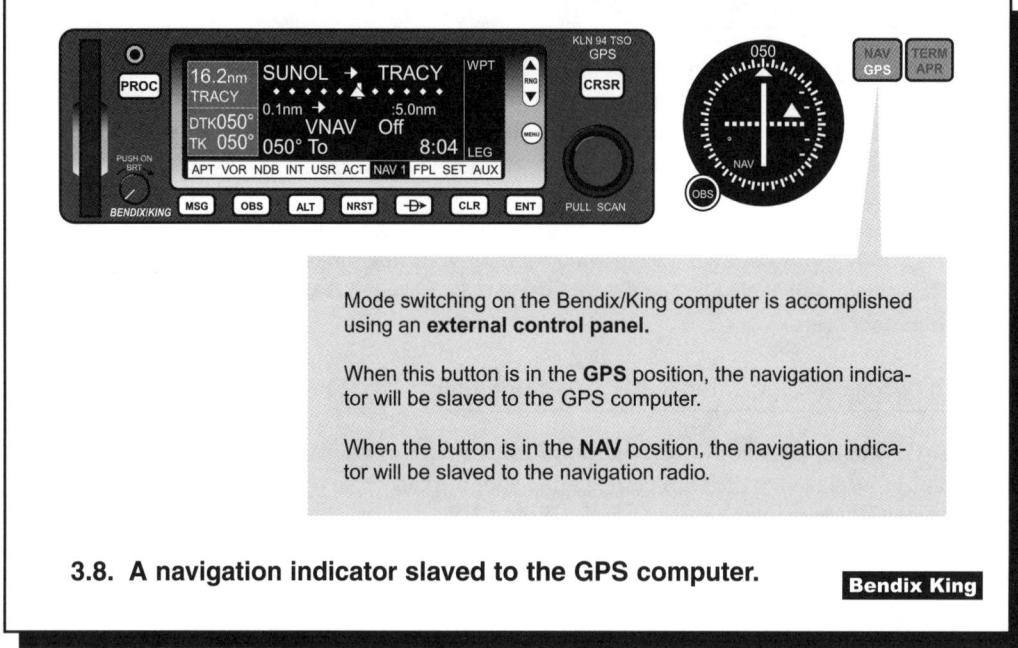

**3.8. A navigation indicator slaved to the GPS computer.**

displayed on the navigation indicator. The consequences of losing track of this important fact can be significant: You may think you are steering along one track when in fact you are steering to a different one. Aside from looking at the annunciation shown on the GPS computer or button, there is an easy way to double-check to make sure your navigation indicator is showing GPS course information. As shown in Figure 3-8, when a navigation indicator is slaved to the GPS computer, you have two course deviation indicators: one on the navigation page on the GPS computer and one on the navigation indicator. Compare them from time to time and make sure they agree.

## SKILL 6: PLAN AND FLY A DESCENT

Most arrival procedures require that you reach an arrival fix at some designated altitude. For example, on your way to Oakdale airport, you might be instructed to cross ECA at 3,000. A directive that requires you to cross a waypoint at a specific altitude is usually referred to as a **crossing restriction.** Descending an aircraft to a given waypoint and altitude not only requires planning but also some precise flying. Let's start by understanding the descent planning problem and solution and see how the GPS computer can assist you.

### Descent Planning Concepts

Planning a descent ultimately requires that you calculate a **top-of-descent point** that takes you from your **cruising altitude** down to the **bottom-of-descent point** that is your crossing restriction. As you know from your instrument flying experience, the top-of-descent point that you choose depends on a combination of your choice of descent speed, the winds, and your choice of rate of descent.

3 / Following the Flight Route

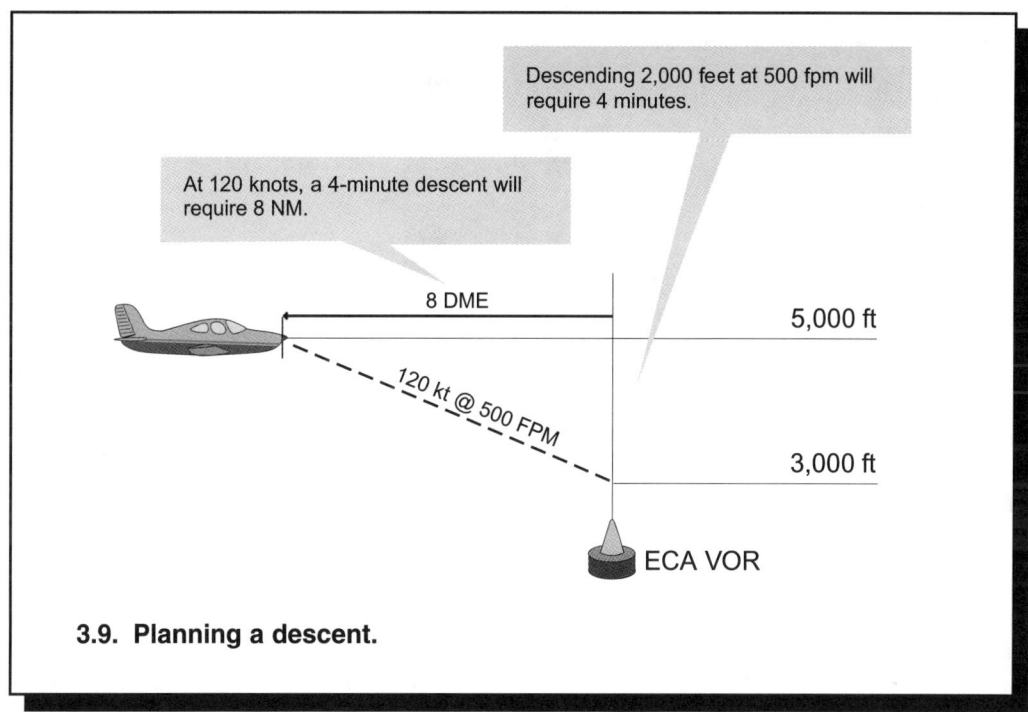

**3.9. Planning a descent.**

For example, consider the scenario illustrated in Figure 3-9 in which you are asked to descend from 5,000 feet and cross ECA at 3,000 feet. If you decide on a descent speed of 120 knots and a descent rate of 500 fpm, you will want to make your top-of-descent point 8 NM prior to ECA (assuming the winds are calm). This should allow you to perform a stabilized descent that delivers you at ECA at 3,000 feet. If you decide on a descent speed of 90 knots, you will need to make your top-of-descent point 6 NM prior to ECA.

This example was simple because we picked easily divisible numbers that allowed you to do quick calculations in your head. When the situation becomes more complex, it is nice to have a computer that can help you with the math.

### Planning a Descent with the Computer

The GPS computer can offer assistance in the laborious task of planning a descent. Building a descent with the computer follows the now familiar process of entering the basics of the descent, letting the computer do the math, and then reviewing what the computer has produced.

Descent planning is accomplished on a separate page using the steps illustrated in Figure 3-10. These steps require you to simply fill in the entries that appear on the page. Note that there is an entry for each of the descent planning concepts we just discussed.

### Descent Flying Concepts

Probably the most important thing to realize about the descent you have just planned is that the descent is really a "wire-in-the-sky." If you start down at the planned top-of-descent point, fly a

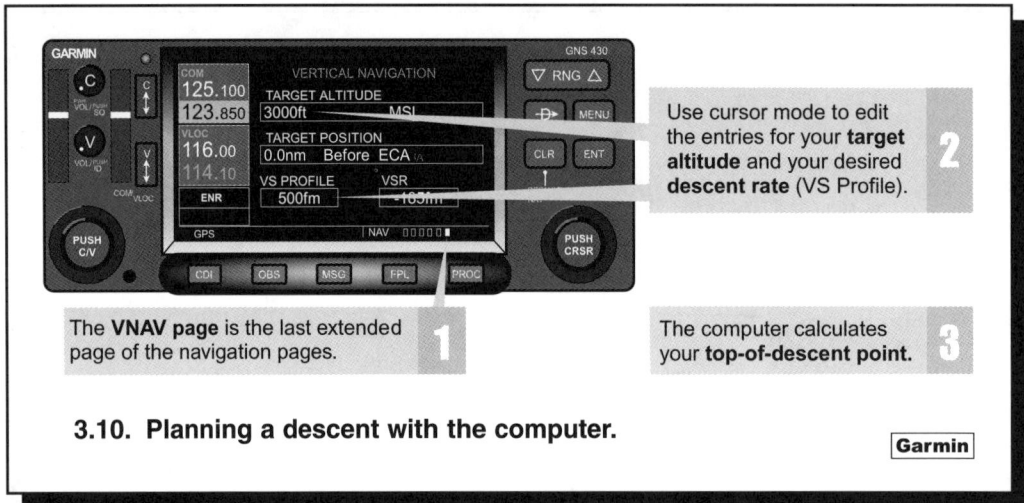

3.10. Planning a descent with the computer.

ground speed of 120 knots, and descend at 500 fpm, you will be flying on a fixed path as if someone had stretched a wire between your top-of-descent point and the bottom-of-descent point. If you maintain your 120-knot and 500-foot-per-minute descent, you will cross a point 6 NM from ECA at exactly 4,500 feet, a point 4 NM from ECA at 4,000 feet, and a point 2 NM from ECA at exactly 3,500 feet. If you find yourself at any different altitude at any of these points, something went wrong, and you can be guaranteed that you will not cross ECA at the required 3,000 feet unless you correct the situation.

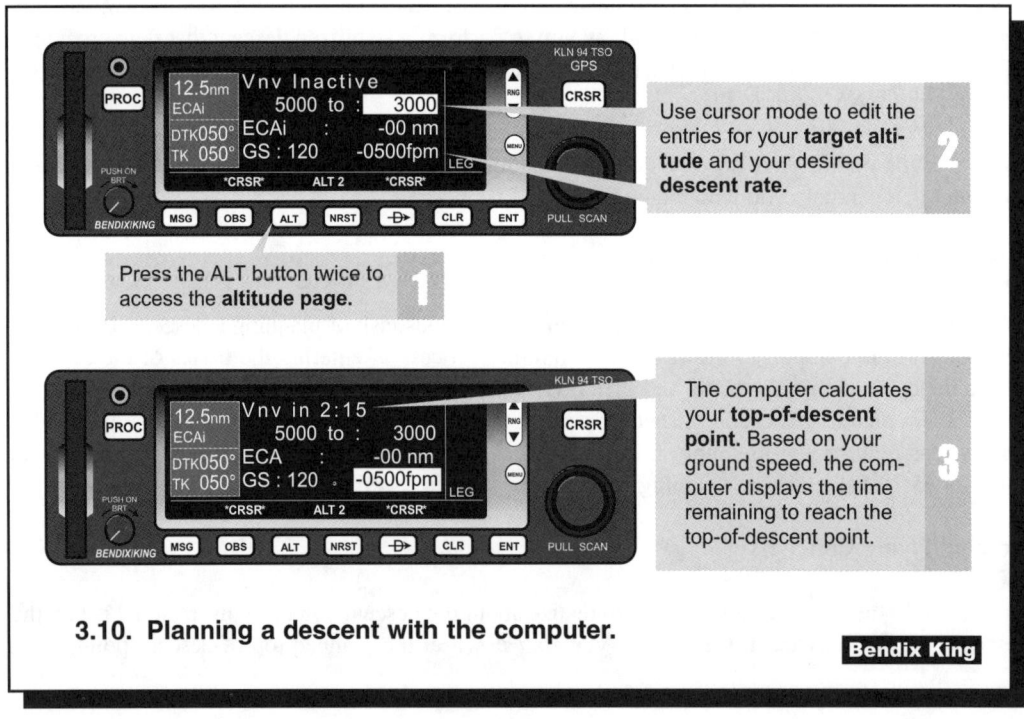

3.10. Planning a descent with the computer.

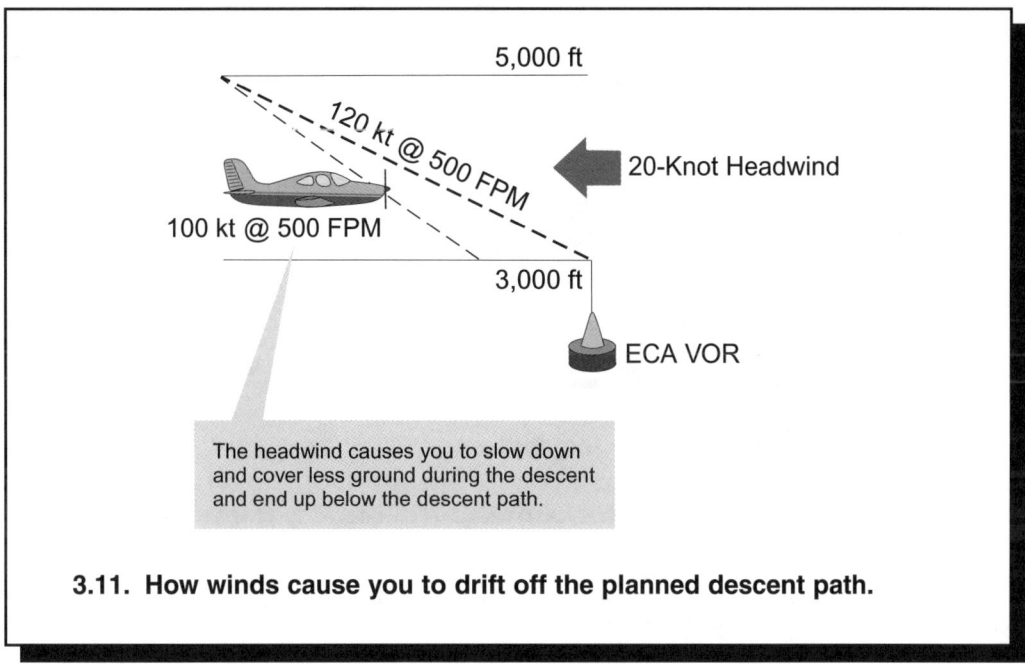

**3.11. How winds cause you to drift off the planned descent path.**

So what can cause you to drift off a planned descent path? You can stray from your planned descent speed, stray from your planned descent rate, or the winds can change unexpectedly and cause your ground speed to change. When this happens, you will drift off the descent path and will need to get back on it. For example, suppose the winds changed to a 20-knot headwind. Your ground speed will now be 100 knots instead of the planned 120 knots. Since you are traveling slower, you will continue to descend at the same rate but will not traverse as much distance, and end up low. The illustration in Figure 3-11 shows how you will have drifted below the planned descent path.

Note that there is nothing wrong with being at this lower altitude. The problem is that you reached the altitude at the wrong place! The winds have slowed your progress and have prevented you from reaching the point where you should be by the time you have descended to this altitude. There are two solutions to this problem: (1) increase your indicated airspeed or (2) decrease your rate of descent.

### Flying a Descent with the Computer

You may have figured out that the real trick in flying a descent is knowing your position relative to the wire-in-the-sky at all times. If you drift off the path, for whatever reason, you know you have to modify your descent with either pitch or power so that you can quickly rejoin the path.

On the Bendix/King GPS computer, you can always tell your position with respect to the descent path by monitoring the display that appears at the top of the **altitude page** (see Figure 3-12). After you pass the top-of-descent point, this line will show the altitude that the aircraft should be currently at if it is to be on the descent path that was calculated. If your altitude matches what is showing on the altitude page, all is going to plan. If your altitude differs from that shown on the altitude page, you have work to do.

The Garmin GPS computer does not display required altitudes during a descent.

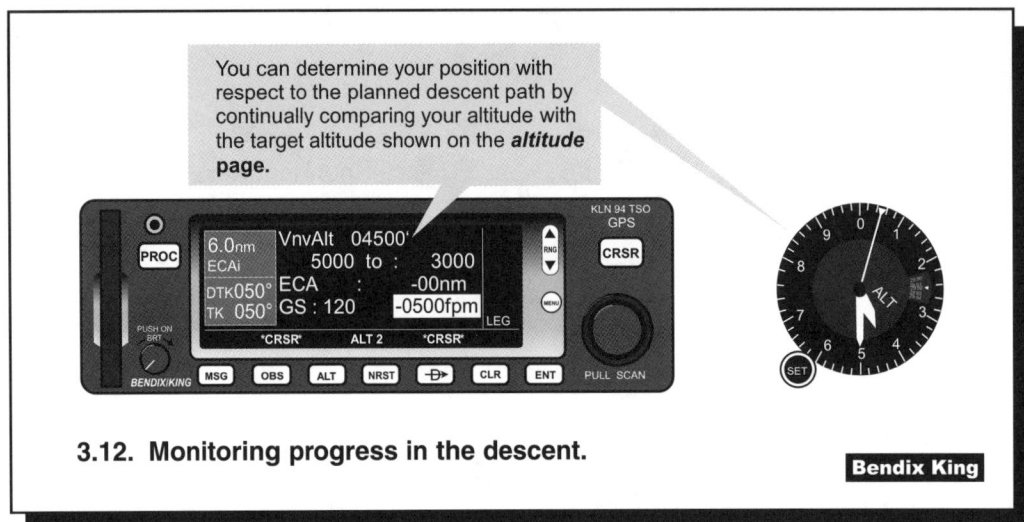

**3.12. Monitoring progress in the descent.**

**Bendix King**

Both GPS computers alert you as you come close to reaching the planned top-of-descent point. These alerts are illustrated in Figure 3-13.

**3.13. Top-of-descent alerts.**

**Garmin**

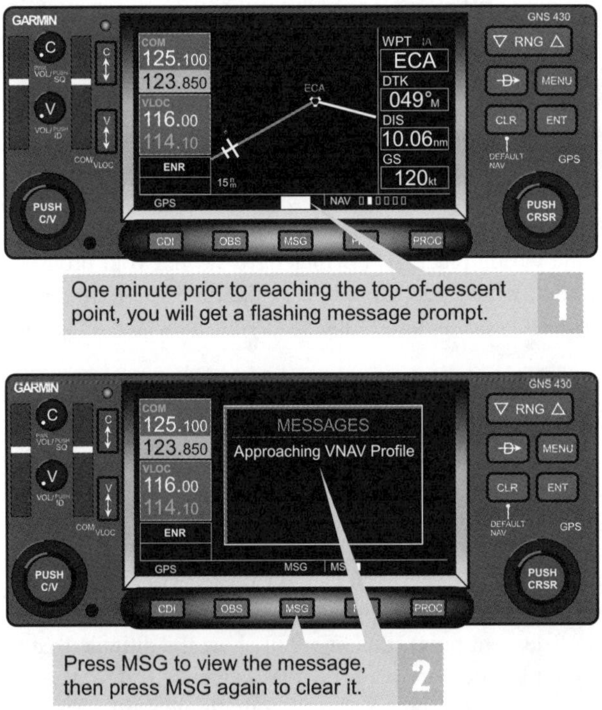

# 3 / Following the Flight Route

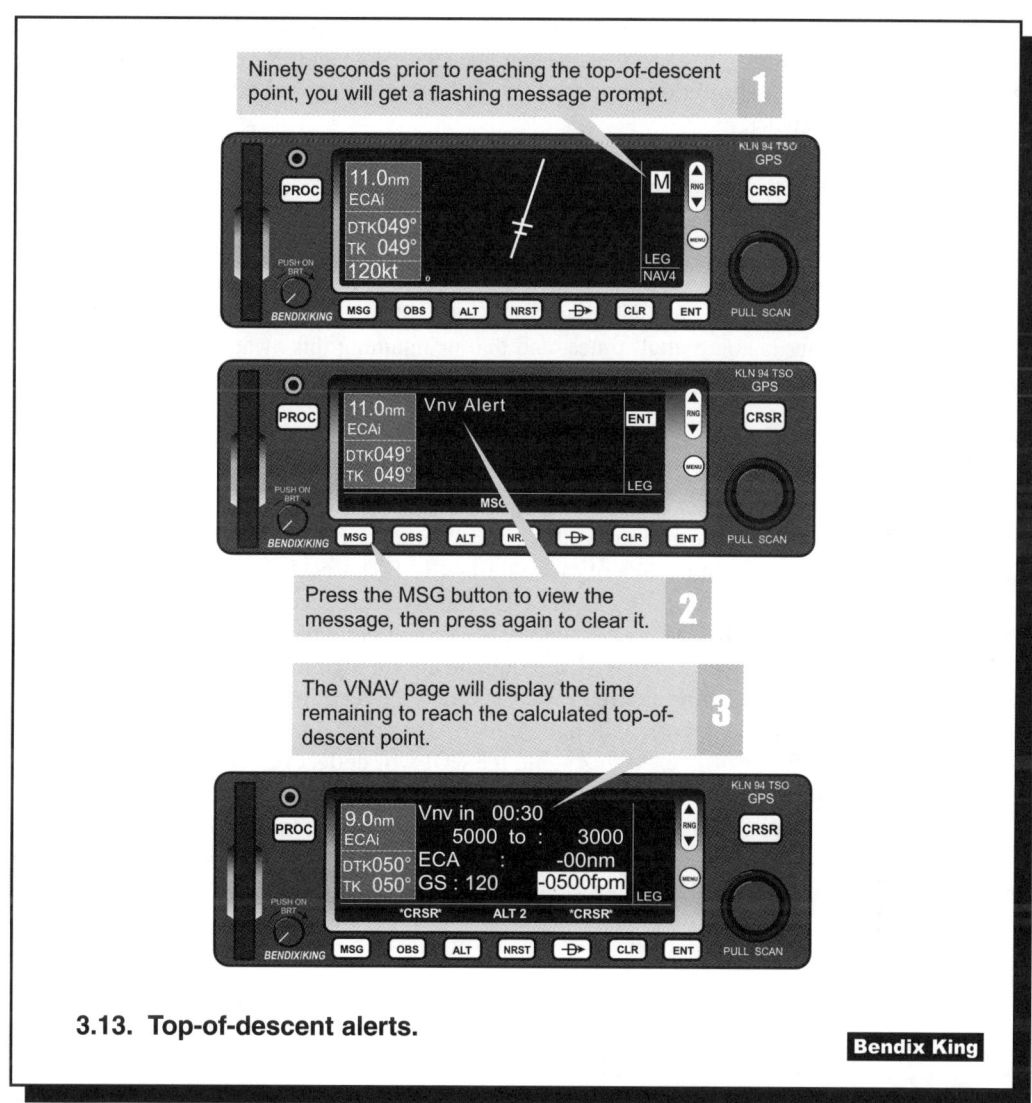

3.13. Top-of-descent alerts.

## SKILL 7: DEMONSTRATE A GPS APPROACH EXPLAINING ALL APPROACH MODE TRANSITIONS

The capabilities of IFR-capable GPS computers make possible a new kind of instrument approach that allows you to use your GPS computer as the principal, and often sole, means of navigation. If you flip through your instrument approach plates, you will see that GPS approaches are becoming increasingly available.

At the present time, there are two kinds of GPS approaches. **GPS overlay approaches** juxtapose a GPS approach procedure over top of an existing VOR, NDB, or RNAV approach procedure. To fly an overlay GPS approach, the aircraft must be equipped with an IFR-certified GPS computer in addition to the other kind of navigation equipment named in the approach. For example, to fly an NDB or GPS approach using GPS, you must be equipped with both an IFR-certified GPS computer and an ADF receiver. **GPS standalone approaches** are instrument approaches that require only an IFR-certified GPS computer.

The simple part of flying a GPS approach is that it is no longer necessary to tune, twist, and identify navigation devices. The waypoints that make up the approach are already listed in your programmed route. Your job is to steer the airplane between the waypoints, with the help of waypoint alerting, turn anticipation, and waypoint sequencing.

The real challenge of flying a GPS approach lies in being prepared for two important changes that happen during the approach. You will find that flying a GPS approach requires an awareness of what the GPS computer is doing at all times, and that maintaining this awareness can be as challenging as steering the airplane.

### From en Route Mode to Terminal Mode

The first important event that happens during your approach occurs once you reach a point within 30 NM of your destination airport. At this critical point during an arrival, regulations require that every GPS computer tighten its integrity monitoring (RAIM). Recall that RAIM is the function that continually checks on the reliability of the GPS signals being received by the computer. The GPS computer performs the RAIM checks automatically and only alerts you in the case that the RAIM requirements are not met. If the RAIM requirements are met, the GPS computer will automatically switch from the **en route mode** to what is called the **terminal mode.** When engaged in the terminal mode, the GPS computer will tighten its RAIM requirements, and increase the sensitivity of the CDI by a factor of five. At this point, the CDI will deflect full-scale when the aircraft drifts 1 NM from the desired track to the active waypoint.

Every GPS announces that it has switched from the en route mode to terminal mode by displaying an annunciation. Figure 3-14 shows the annunciations for both GPS computers. It is your responsibility to monitor for RAIM failure annunciations and the appearance of the terminal mode annunciation.

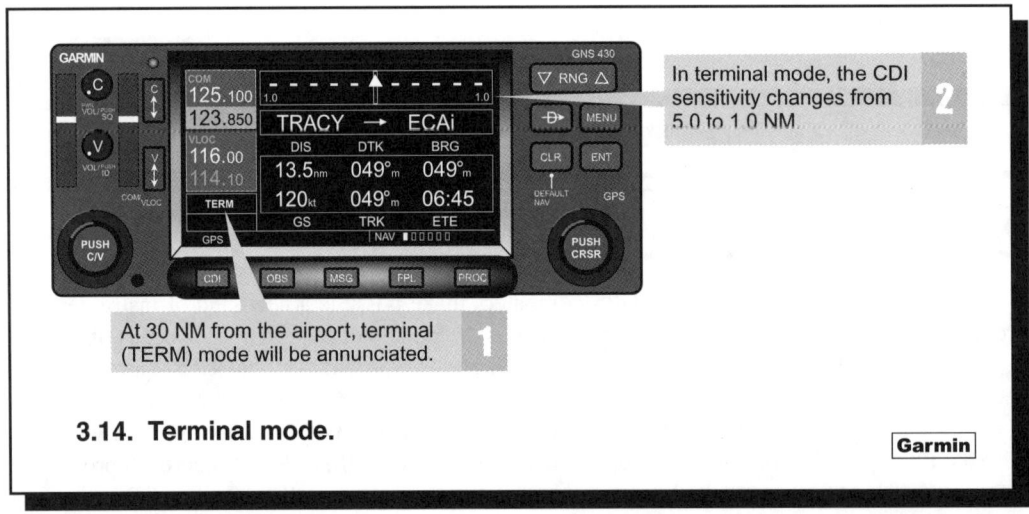

At 30 NM from the airport, terminal (TERM) mode will be annunciated.

In terminal mode, the CDI sensitivity changes from 5.0 to 1.0 NM.

**3.14. Terminal mode.**

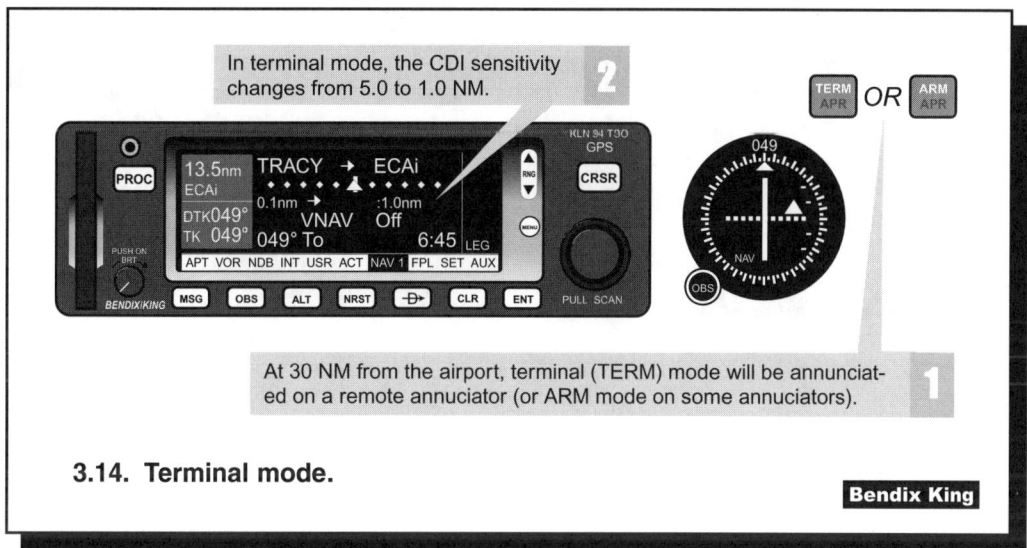

**3.14. Terminal mode.**

## From Terminal Mode to Approach Mode

The second important event occurs 2 NM prior to reaching the final approach fix (FAF). At this point, the GPS computer should automatically switch from the terminal mode to the **approach mode.** When engaged in the approach mode, the GPS computer will again tighten its RAIM requirements, and increase the sensitivity of the CDI even further. A full-scale deflection of the CDI will now occur if you drift 0.3 NM from the desired track to the active waypoint.

The annunciations for the switch to approach mode are shown in Figure 3-15. As long as the annunciation for approach mode is showing, you may continue the approach. If, for any reason, the computer fails to switch to approach mode, or the approach mode annunciation disappears, the regulations require you to fly the published missed approach procedure. You are not authorized to descend to the minimum descent altitude (MDA).

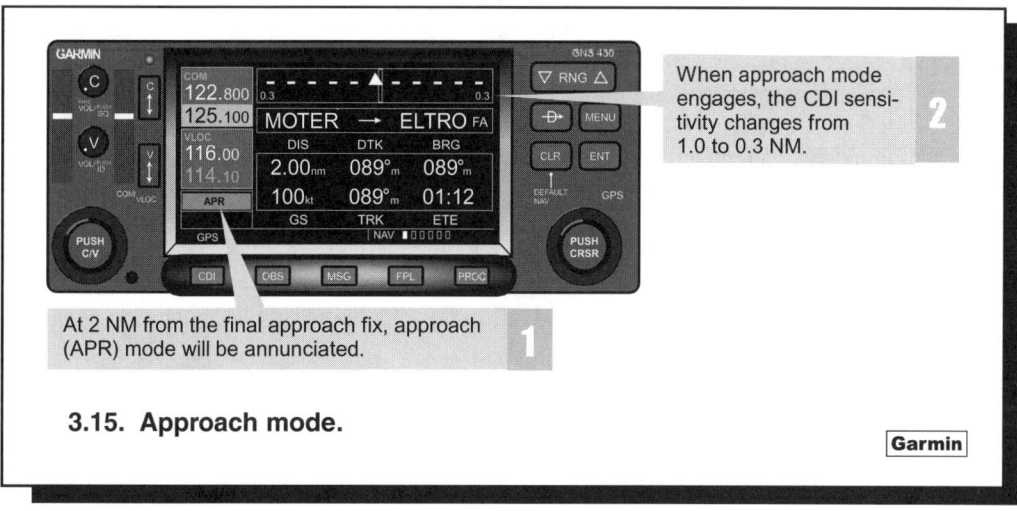

**3.15. Approach mode.**

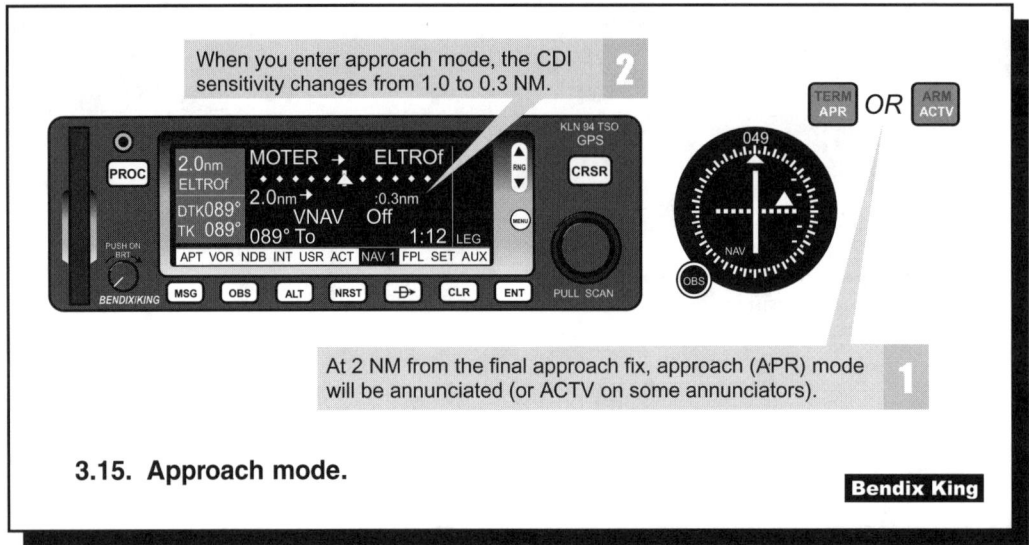

3.15. Approach mode.

After switching to approach mode 2 NM short of the FAF, you will have about a minute to ready yourself for the final approach segment. This is a good time to take care of any checklist items that need to be accomplished. The switch to the approach mode is also a good time to recheck that the navigation indicator is indeed slaved to the GPS. Checking both of these indications at 2 NM prior to the FAF should be a checklist item for every GPS approach.

It is important to note that no further changes to the GPS computer should be made once the 2-NM point has been reached, since these will likely result in the cancellation of the approach mode.

A common error committed by many students is to mistake the 2-NM approach mode point for the FAF. Remember, you still have 2 NM to travel before you reach the FAF and can begin your descent down to the MDA.

## More on Mode Awareness

Throughout this chapter, we have discussed the topic of modes and mode awareness many times. In each case, the importance of remaining aware of which modes are engaged at all times has been emphasized. For many reasons, the task of maintaining awareness of which modes are engaged is a consistent problem area for pilots learning to use cockpit automation. Every pilot learning to use a GPS computer is encouraged to pay particular attention to the mode awareness task. You may want to think about augmenting your checklists to include checks for the mode checks that we discussed in this chapter. Four important mode checks during a GPS approach are as follows:

1. Verify the terminal mode 30 NM from the airport.
2. Verify that the computer is set to sequencing mode.
3. Verify that the navigation indicator is set to reflect the GPS course.
4. Verify the approach mode 2 NM from the FAF.

Use regular call-outs that encourage monitoring for these important events. Note that the consequences of losing awareness of any of these modes can be quite severe, introducing the very real possibility of controlled flight into terrain.

# 3 / Following the Flight Route

 **PRACTICE SESSION (AIRPLANE AND/OR SIMULATOR)**

Now you know how to build and follow a flight route using your GPS computer. Even if you plan to practice your new skills in an airplane, take some time to run through the KSJC-O27 flight covered in the book using your simulator. This way you can refer back to the book when you encounter something that doesn't make sense. Using a simulator also allows you to stop and think for a while on your first few times through.

After you have mastered the KSJC-O27 flight, try some other ones with the simulator or in the airplane. Pick a destination airport that has a GPS approach. Pick a GPS approach that does not require a procedure turn. Fly under VFR and use the GPS computer as your principal means of navigation during the entire flight. If you are using an airplane for your flights, stick with simulated instrument conditions for now, unless you are flying with an instrument CFI who is already proficient with GPS computers. When you're not quite an expert yet, things can fall apart pretty quickly in the clouds.

### SKILL 4: Ensure Adequate GPS Signal Reception

1. Point out that no "RAIM Not Available" messages are showing at important times during your flight, such as prior to departure or prior to commencing a GPS approach.

### SKILL 5: Demonstrate How the GPS Computer Provides Lateral Guidance Along the Route

1. Point out which GPS computer pages are most useful for following your progress during the flight.
2. Demonstrate position awareness by announcing your arrival at each waypoint along the route and making regular call-outs for the distance and track to the active waypoint.
3. Point out waypoint alerting as you approach each waypoint.
4. Announce the beginning of turn anticipation as you approach each waypoint in your route.
5. Point out the sequencing of each active waypoint and show that the computer is engaged in the sequencing mode.

### SKILL 6: Plan and Fly a Descent

1. Build a descent to cross a designated waypoint along the route at a specified altitude. Cross ECA at 3,000 for the KSJC-O27 route.
2. Cross the designated waypoint at the specified altitude ± 100 feet.

## SKILL 7: Demonstrate a GPS Approach Explaining All Approach Mode Transitions

1. Call out the engagement of the terminal mode and point out the change in CDI sensitivity.
2. Begin the approach at the published IAF.
3. Call out the engagement of the approach mode at 2 NM short of the FAF and show that the navigation indicator is slaved to the GPS computer.
4. Fly the approach to the MDA and then perform a straight-in or circle-to-land procedure.

### How Did You Do? (Common Errors)

At the end of your flight, think about each of the skills described above and assess yourself on how well you did. I ran dozens of student pilots and CFIs through these exercises and found that the most common difficulty was remembering to call out mode changes during the approach. This "mode awareness" skill is new for most pilots, and it takes a little practice to work it into your routine. If you didn't do so well this flight, think about it and give it another try during your next flight.

### DVD DEMONSTRATION DISC

Skills 4, 5, 6, and 7 are demonstrated using the Garmin GNS 30 on the accompanying DVD disc.

CHAPTER 4

# Modifying the Flight Route

*This chapter teaches you how to make simple modifications to the flight route you have programmed into the GPS computer. By the end of this chapter, you will have mastered four new skills:*

SKILL 8: *Demonstrate the Direct-To function.*
SKILL 9: *Add and delete waypoints from your flight route.*
SKILL 10: *Select a different approach or transition.*
SKILL 11: *Demonstrate an emergency diversion.*

Part of the challenge of using cockpit automation is dealing with modifications to your planned flight route that are given to you by air traffic control (ATC). In this chapter, you will learn about four simple route modifications that can be made to your route in-flight. Mastering these skills on the ground will allow you to perform en route modifications to the route stored in your GPS computer with a minimum of head-down time.

## SKILL 8: DEMONSTRATE THE DIRECT-TO FUNCTION

The simplest kind of modification issued by ATC is one that is usually requested by pilots. This modification requires that you proceed "direct to" a waypoint that occurs later in the flight plan.
Suppose you are en route to Oakdale and just approaching Sunol intersection and you receive a clearance to proceed direct to ECA. You can accomplish this simple modification using the technique illustrated in Figure 4-1.
After you complete the Direct-To operation, ECA becomes the active waypoint. It is important to be aware of the new track that the computer has calculated. The computer has created a new desired track from the point at which you made the Direct-To selection to the waypoint named in the

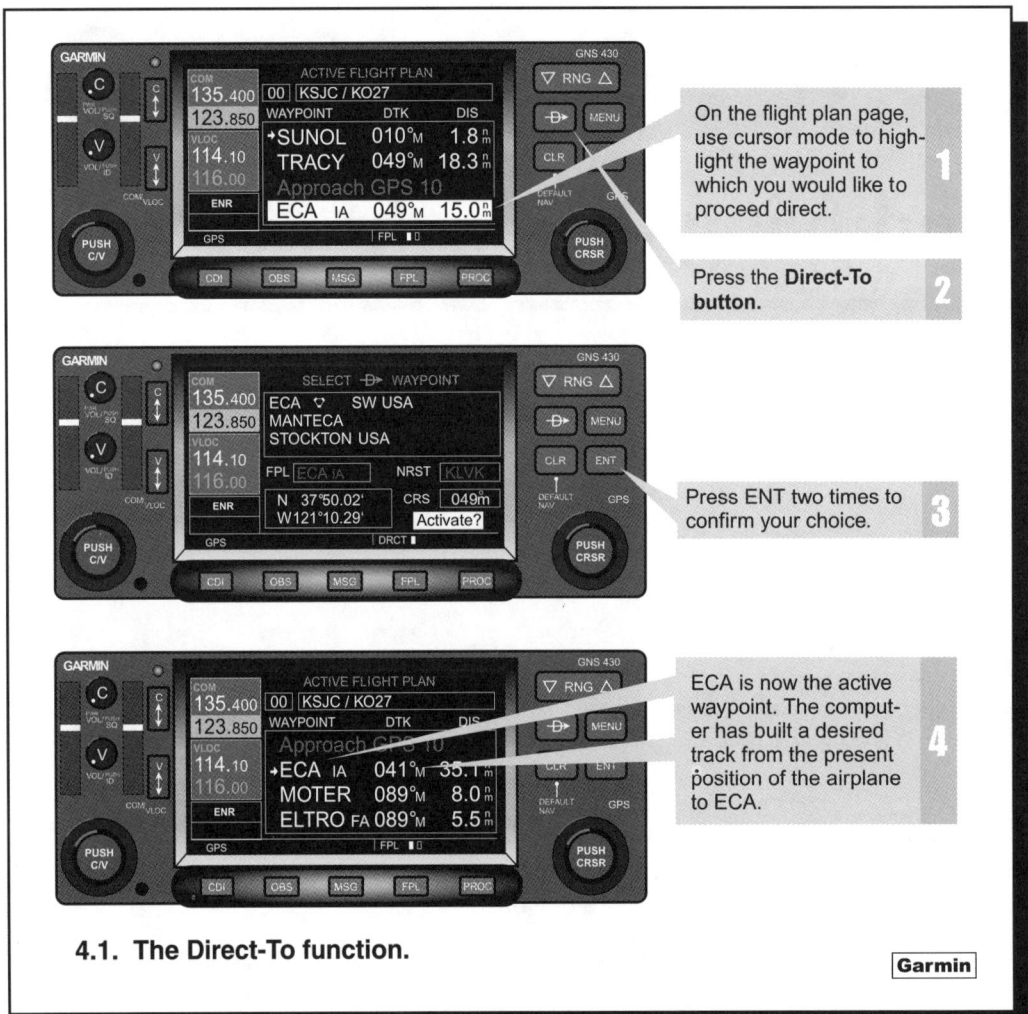

4.1. The Direct-To function.

Direct-To operation. There is no longer any reference to the desired track between the previous waypoint and Sunol. It is never a good idea to ask for direct clearances without double-checking to see that your new direct route is indeed clear of all significant terrain.

4 / Modifying the Flight Route 63

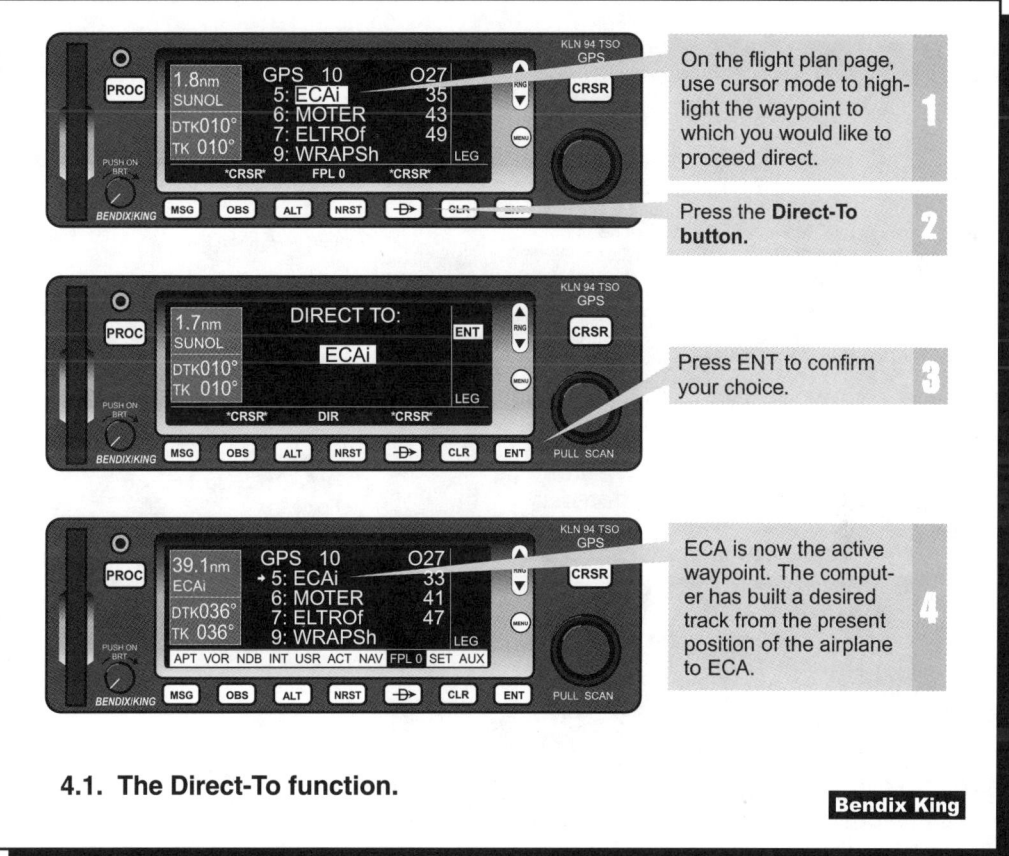

4.1. The Direct-To function.

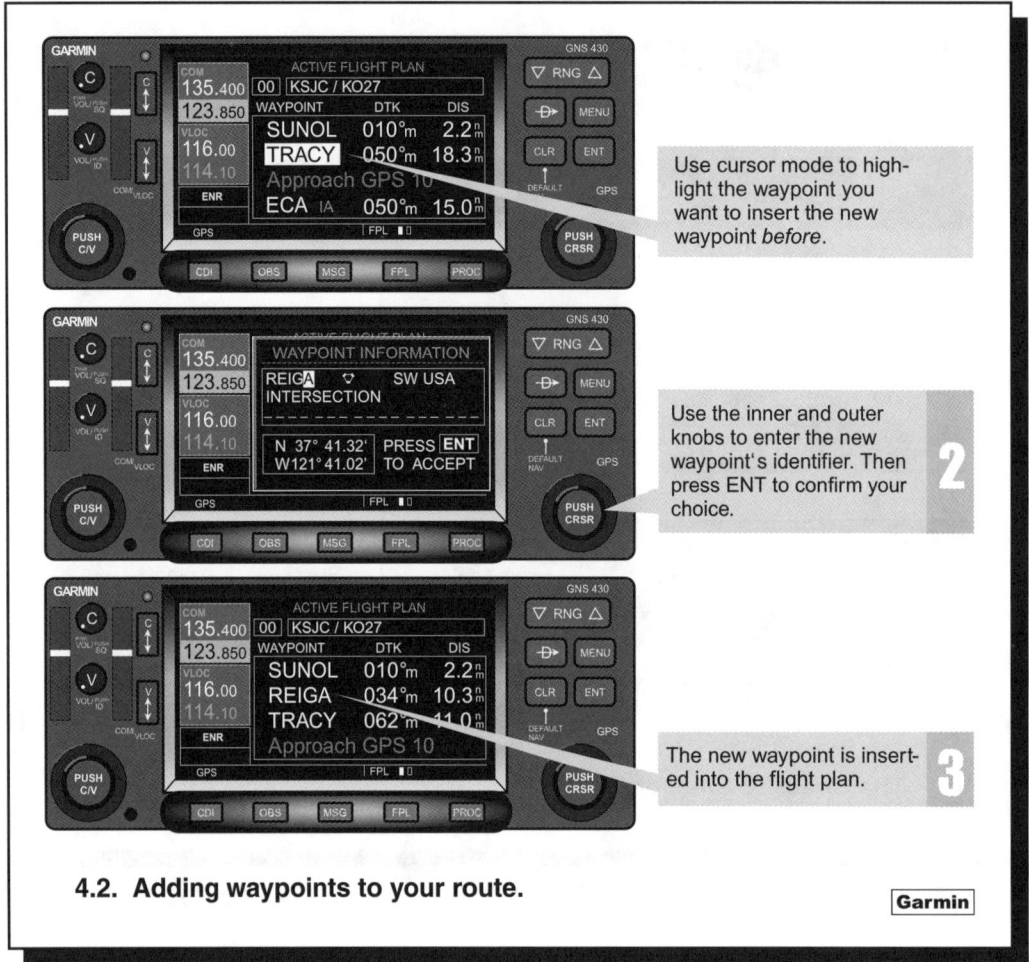

4.2. Adding waypoints to your route.

## SKILL 9: ADD AND DELETE WAYPOINTS FROM YOUR ROUTE

Waypoints can be added to the route on the flight plan page using the technique illustrated in Figure 4-2. Deleting waypoints can be accomplished using the simple process shown in Figure 4-3.

## SKILL 10: SELECT A DIFFERENT APPROACH OR TRANSITION

Suppose you are en route to Oakdale when you are instructed to use Moter as your initial approach fix instead of ECA. This modification can be made following the steps shown in Figure 4-4.

# 4 / Modifying the Flight Route

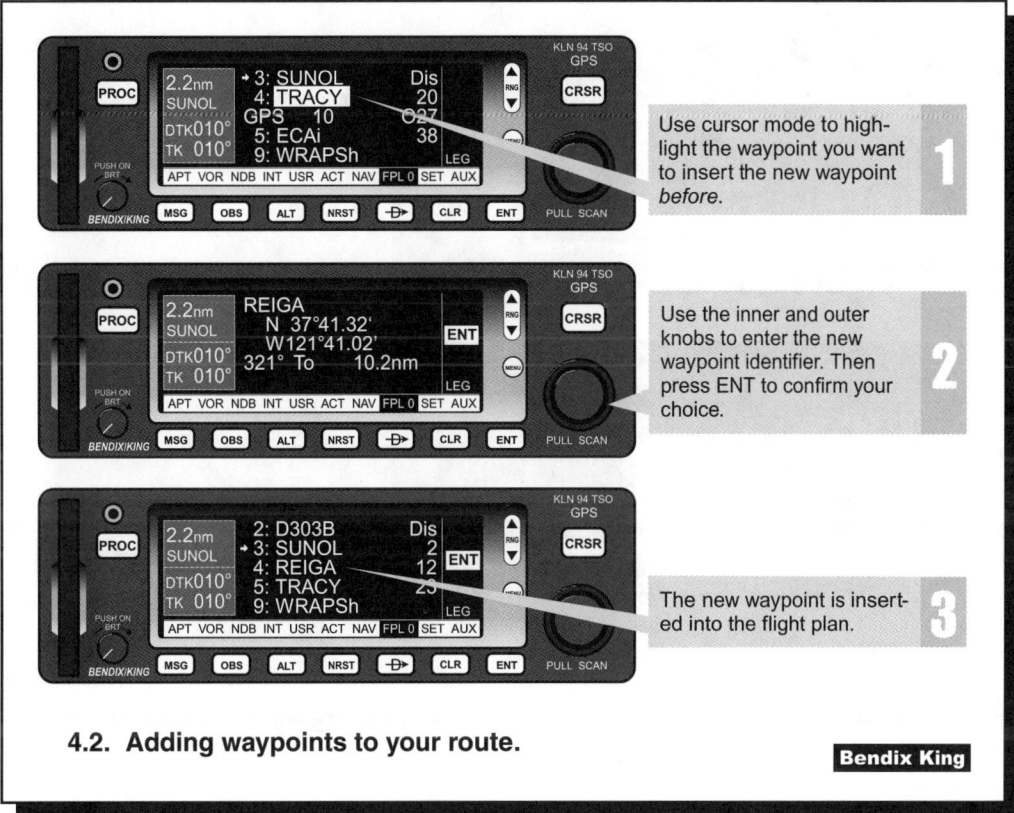

4.2. Adding waypoints to your route.

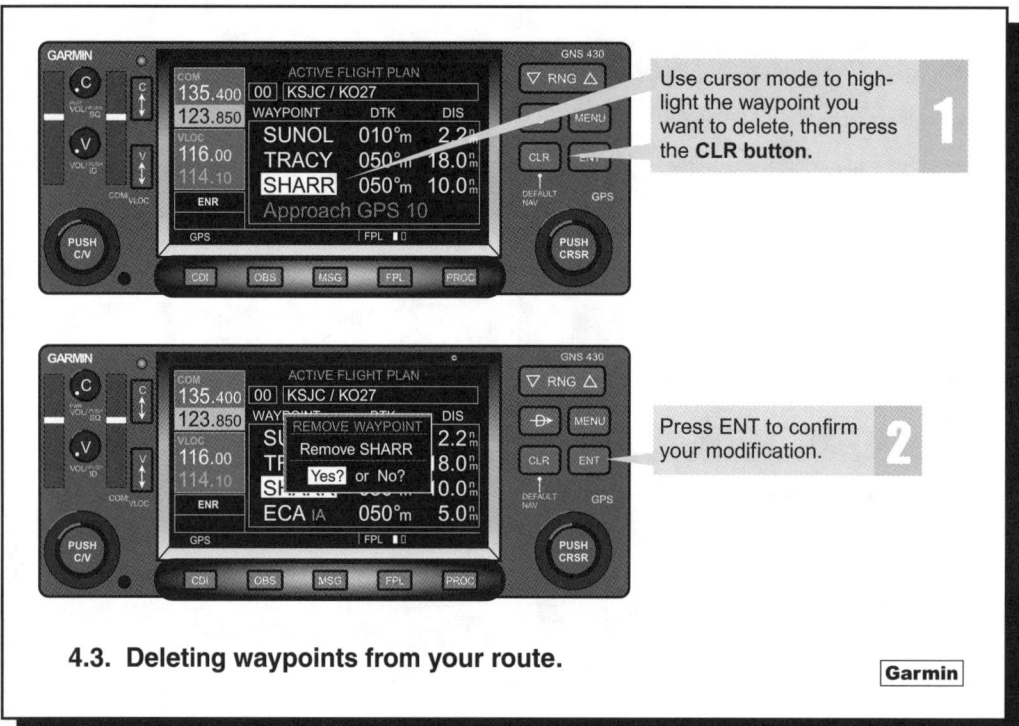

4.3. Deleting waypoints from your route.

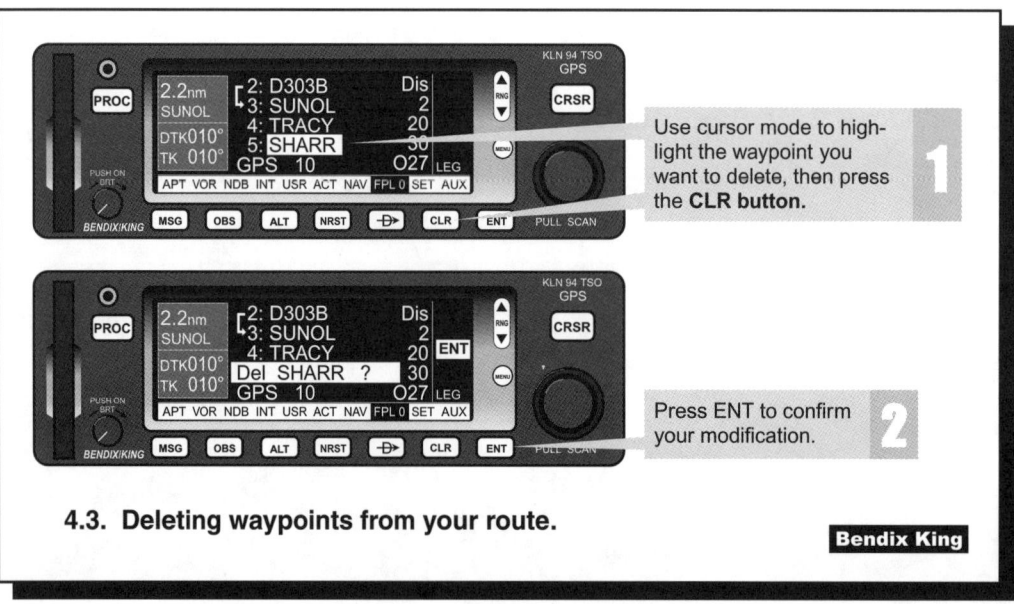

4.3. Deleting waypoints from your route.

4 / Modifying the Flight Route    67

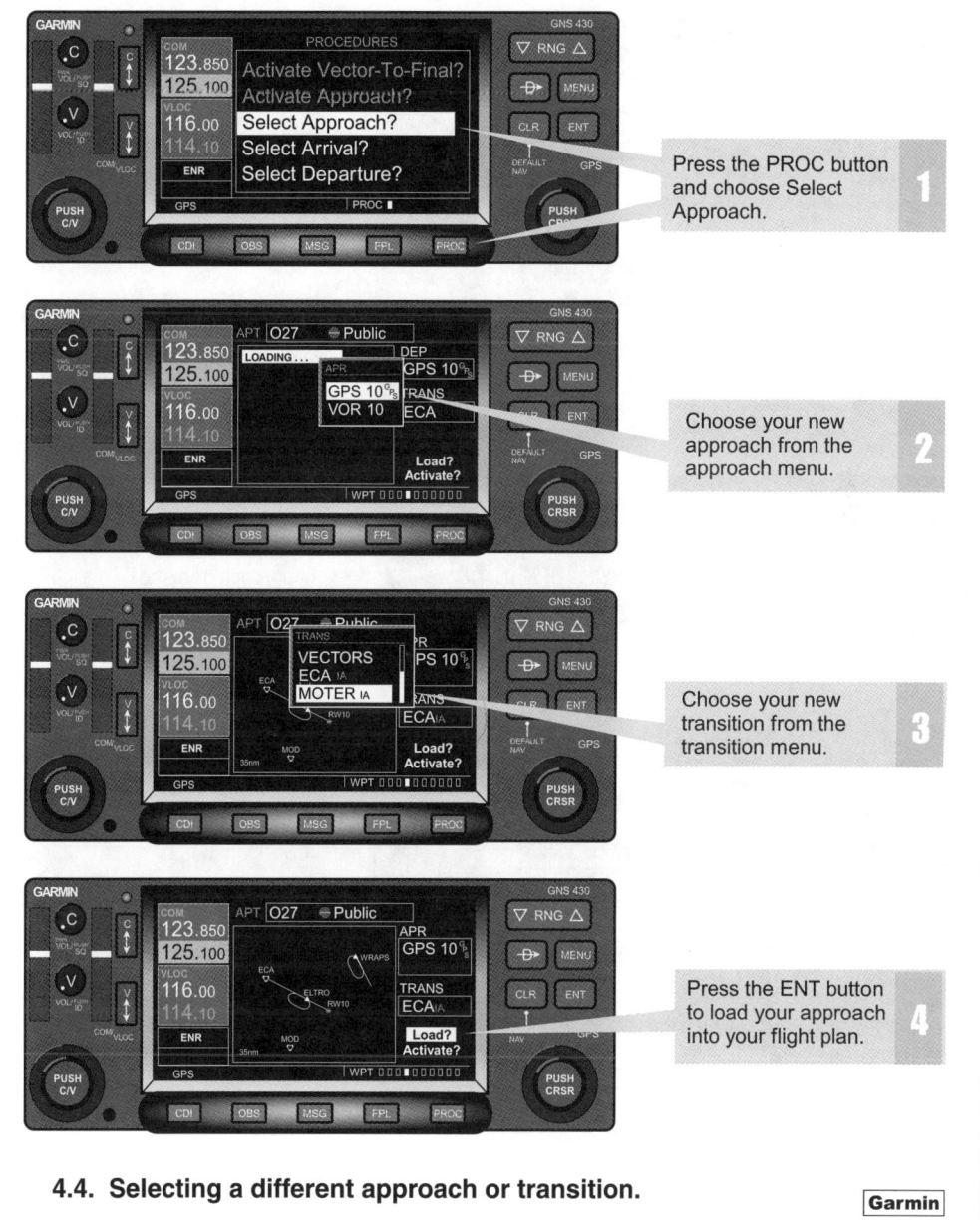

**4.4. Selecting a different approach or transition.**

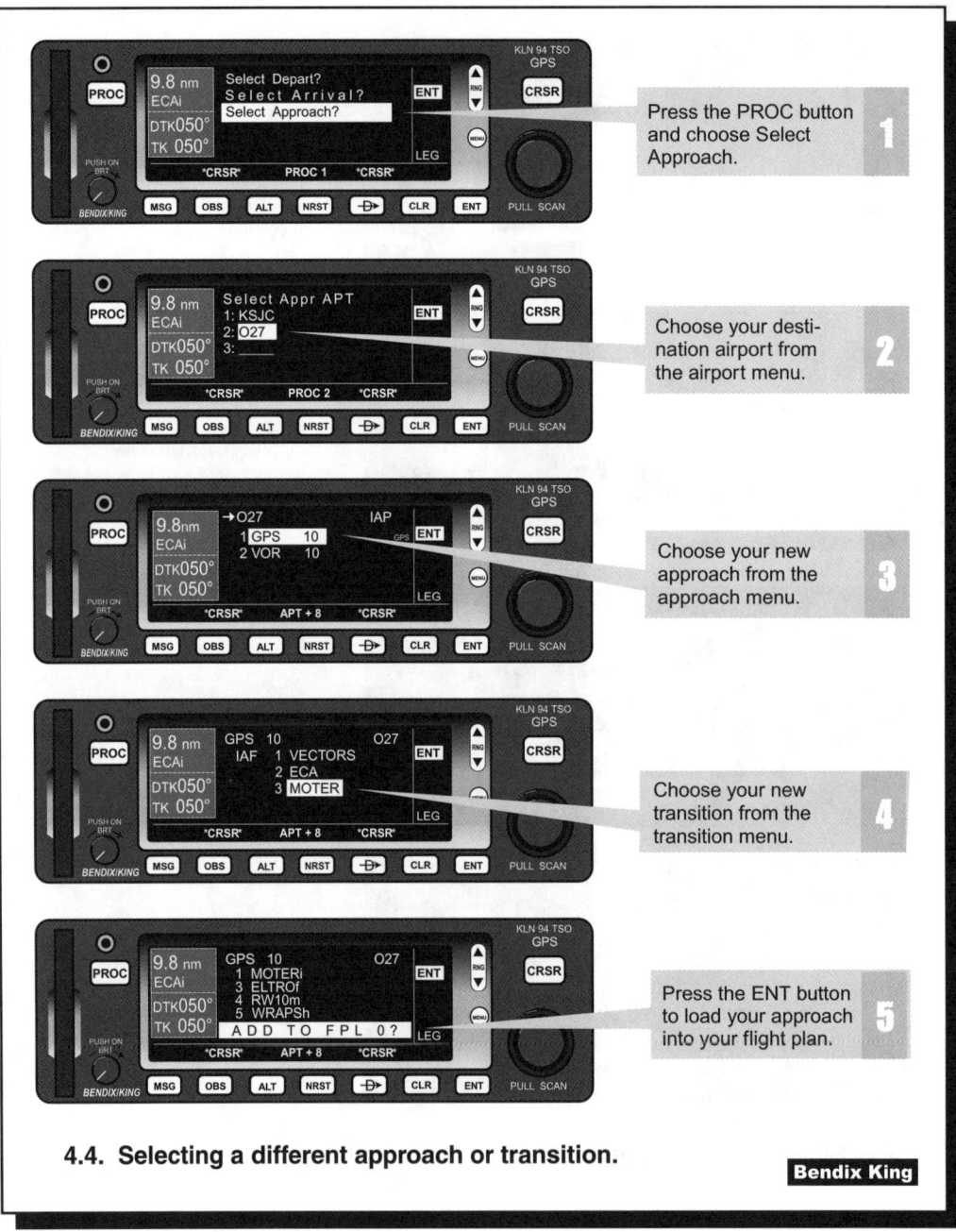

4.4. Selecting a different approach or transition.

# 4 / Modifying the Flight Route

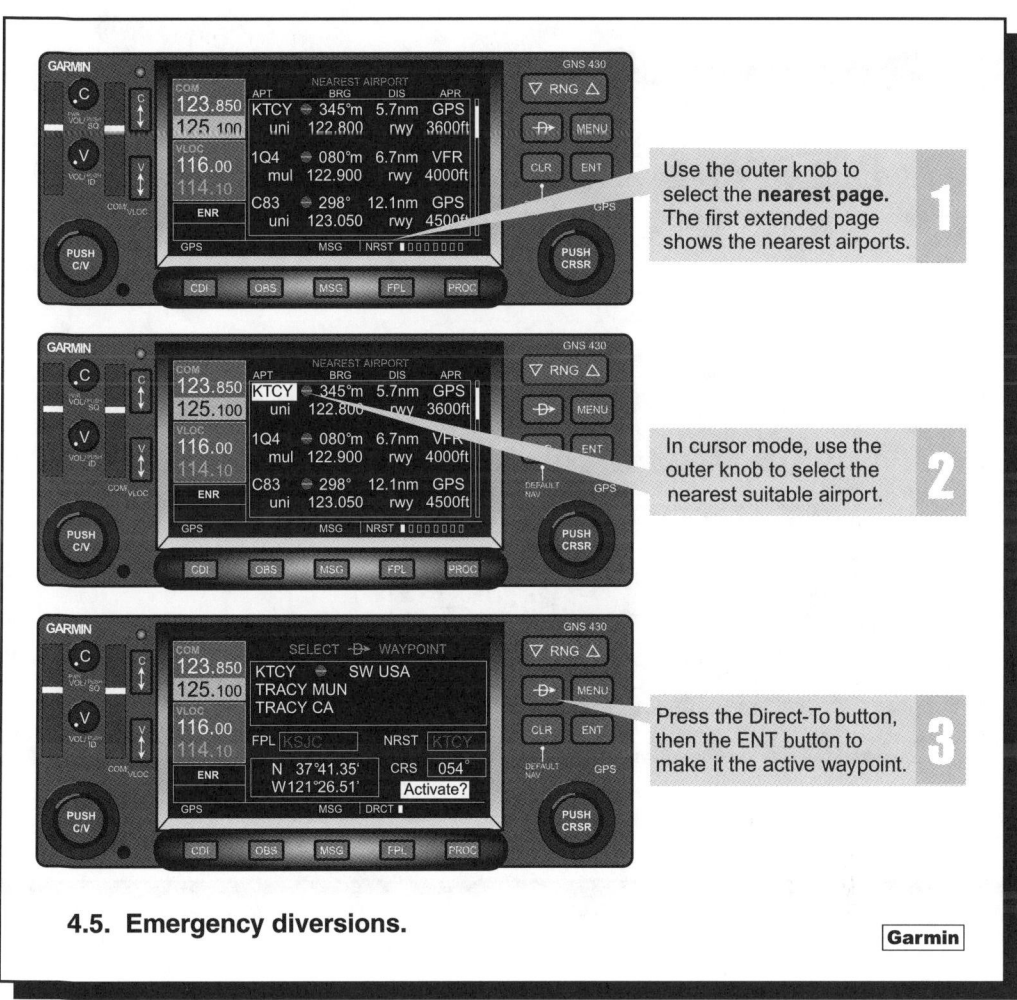

4.5. Emergency diversions.

## SKILL 11: DEMONSTRATE AN EMERGENCY DIVERSION

One of the most useful features of a GPS computer is its ability to provide you with immediate access to a large navigation database. Consider the case of an emergency in which you have to quickly locate a suitable airport for an immediate diversion. Figure 4-5 shows you how to find a suitable alternate airport with just a few button presses.

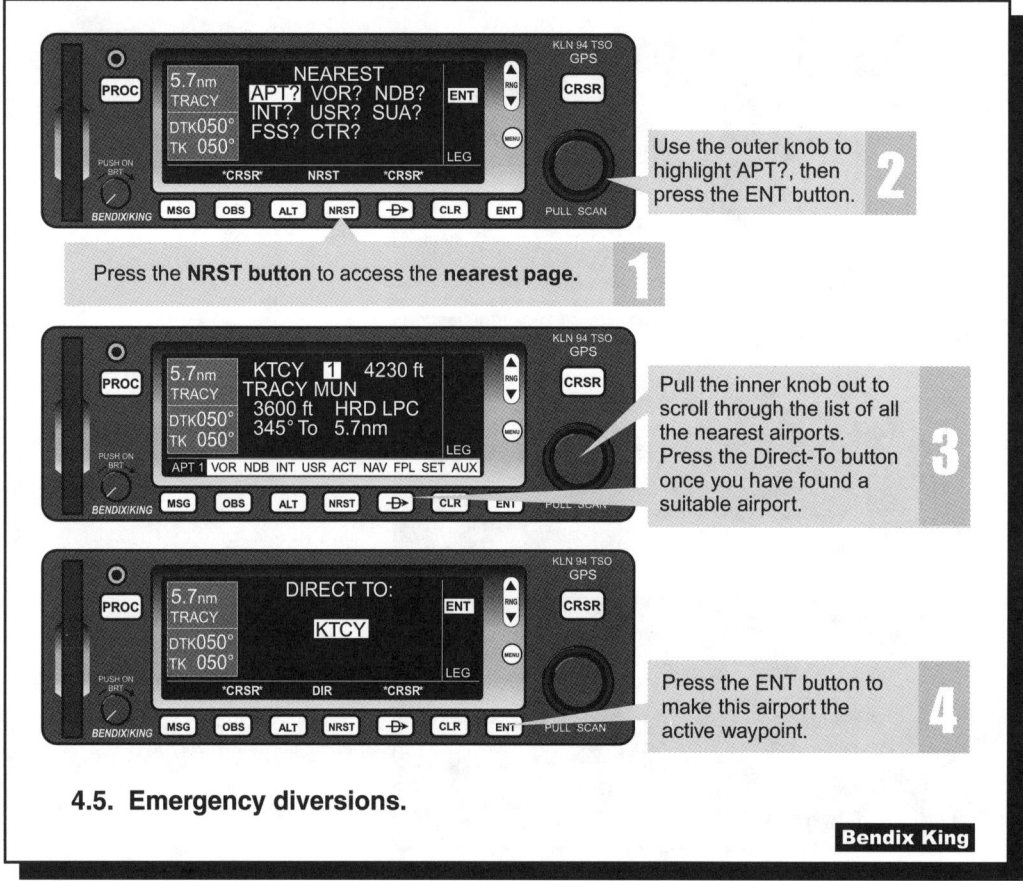

4.5. Emergency diversions.

## 4 / Modifying the Flight Route

 **PRACTICE SESSION (AIRPLANE AND/OR SIMULATOR)**

Using your simulator or an airplane, take a few minutes to practice the four new skills you have learned.

### SKILL 8: Demonstrate the Direct-To Function

1. Choose waypoints that appear in your route and program the GPS computer to proceed directly to them.

### SKILL 9: Add and Delete Waypoints from Your Route

1. Practice adding and deleting waypoints from the flight route.

### SKILL 10: Select a Different Approach or Transition

1. Install an approach at any airport and then select a different approach or transition.

### SKILL 11: Demonstrate an Emergency Diversion

Practice a few emergency diversions.

1. Use the GPS computer to choose an appropriate airport for landing.
2. Program Direct To the chosen airport.
3. Identify which runway and communications frequency will be used at the chosen airport.

### How Did You Do? (Common Errors)

The skill that most pilots had trouble with for this chapter was the emergency diversion. Pilots often forgot that the most important step was choosing a place to go and getting the airplane pointed at it. Including these new button pushing skills into your existing emergency procedures requires some prior thought about what is most important when the pressure is on.

 **DVD DEMONSTRATION DISC**

Skills 8, 9, 10, and 11 are demonstrated using the Garmin GNS 430 on the accompanying DVD disc.

# CHAPTER 5

# Advanced Maneuvers

*This chapter teaches you how to deal with more sophisticated modifications to your programmed flight route. Each of these modifications requires you to fly off your programmed route and to then rejoin your route at some later point. By the end of this chapter, you will have mastered six new skills:*

> SKILL 12: Intercept an inbound course to the active waypoint.
> SKILL 13: Intercept an inbound course to a different active waypoint.
> SKILL 14: Holds.
> SKILL 15: Procedure turns.
> SKILL 16: Missed approaches.
> SKILL 17: Explain the importance of continuously monitoring the GPS computer.

During your first flight, you probably noticed how the GPS computer simplifies the task of flying from one waypoint to the next in your planned route. Instead of tuning, twisting, identifying, and wondering where you are, the GPS seems to almost magically guide you along your route. It seems that as long as you keep the CDI centered, you are on course.

What makes the navigation task so simple is the highly automated sequencing mode that you have used exclusively up until now. The sequencing mode makes the route following task simple by providing these two useful services for you:

1. **Waypoint Sequencing:** As you fly toward and reach each waypoint in your route, the sequencing mode automatically advances the active waypoint to the next waypoint in the route.
2. **Desired Tracks:** The sequencing mode always steers you along the desired track to the active waypoint.

In other words, when you use the sequencing mode, you never have to figure out where you are going next or how you are going to get there. Throw in a display that shows you where you are now, and it might seem that getting lost is almost impossible.

## The Nonsequencing Mode

It turns out that every IFR-capable GPS computer offers a second mode of operation that allows you to do things the hard way. This mode does not perform waypoint sequencing, and it does not automatically steer you along the desired tracks that are computed by the GPS computer. For this reason, this mode is often referred to as the **nonsequencing mode.** Both Garmin and Bendix/King use the name **OBS mode** for the nonsequencing mode.

You might wonder why anyone would choose to use this more labor-intensive mode when the sequencing mode is available. In this chapter, you will learn about five everyday situations in which the automated features of the sequencing mode are undesirable and in which you will be thankful for having a way to turn them off. All five of these situations have one thing in common. Instead of flying your route from waypoint to waypoint as planned, each situation is one in which ATC instructs you to depart from your originally planned route and then rejoin it later. You will see that the nonsequencing mode will allow you to temporarily stop following your planned route, do whatever ATC has instructed you to do, then later rejoin your planned route and reengage the sequencing mode.

## SKILL 12: INTERCEPT AN INBOUND COURSE TO THE ACTIVE WAYPOINT

Let's consider a common situation in which you do not want to follow the desired tracks between the waypoints that have been calculated by your GPS computer. Consider the Sunol Five departure procedure you have been asked to follow out of San Jose International Airport. Figure 5-1 diagrams the procedure.

Let's compare the diagram of the Sunol Five departure procedure with the series of waypoints that have been inserted in your route in the GPS computer after you have selected the Sunol Five departure procedure. The waypoints for the procedure are shown on the flight plan pages in Figure 5-2.

Departing San Jose airport, the active waypoint is named D303B and appears to be located about 2 NM from the runway on a bearing of approximately 303 degrees. This waypoint seems to nicely correspond to the 1.8 DME fix that the Sunol Five procedure requires you to cross. In fact, the D303B waypoint was created by the database authors to represent the 1.8 DME position. The D in D303B stands for DME. The 303 represents a 303-degree magnetic course from the SJC VOR. The B stands for 2 NM (A means 1 NM, C means 3 NM, and so on).

Looking at the third waypoint in the GPS computer route, you can see that you have a problem. The GPS computer plans to guide you from the 1.8 DME fix directly to Sunol intersection. This is not what you have been instructed to do. The GPS route seems to be missing the part about the 040-degree heading and intercepting the 009-degree course to Sunol from the west side of the course.

What you really want to do is to put the GPS route aside for a few minutes and fly the assigned heading of 040 degrees. Once established on that heading, you would then like to proceed to Sunol intersection, only by intercepting and following a course that you choose, rather than the direct course that the GPS computer has chosen. This is precisely the sort of thing that the nonsequencing (OBS) mode allows you to do.

# 5 / Advanced Maneuvers

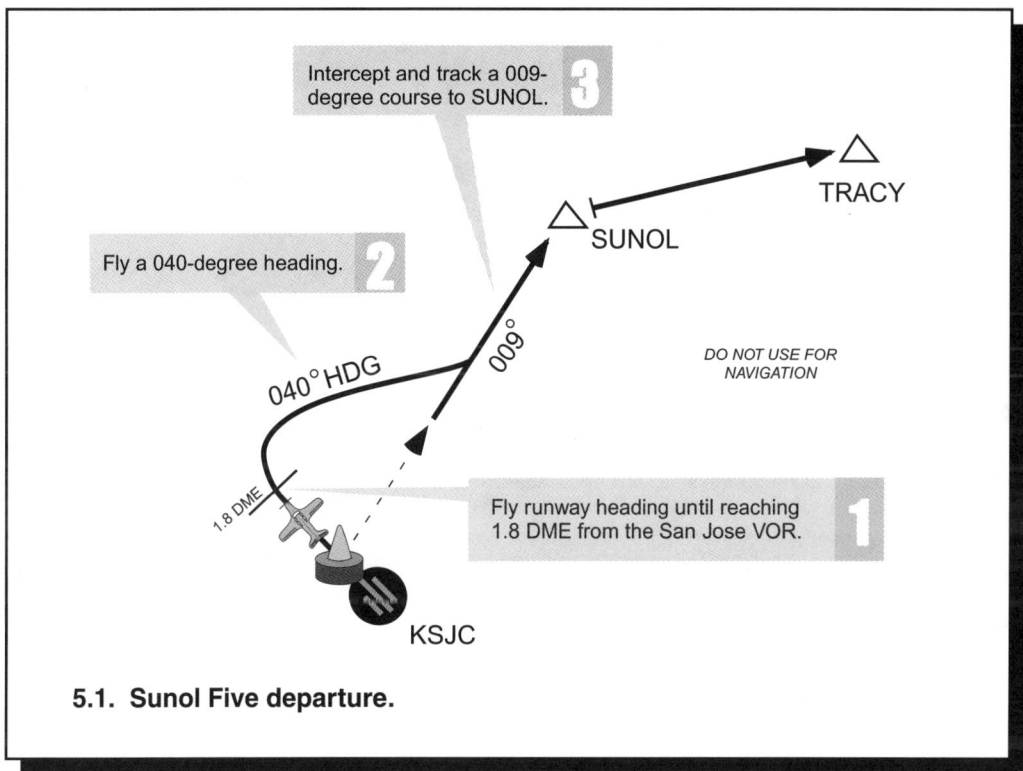

5.1. Sunol Five departure.

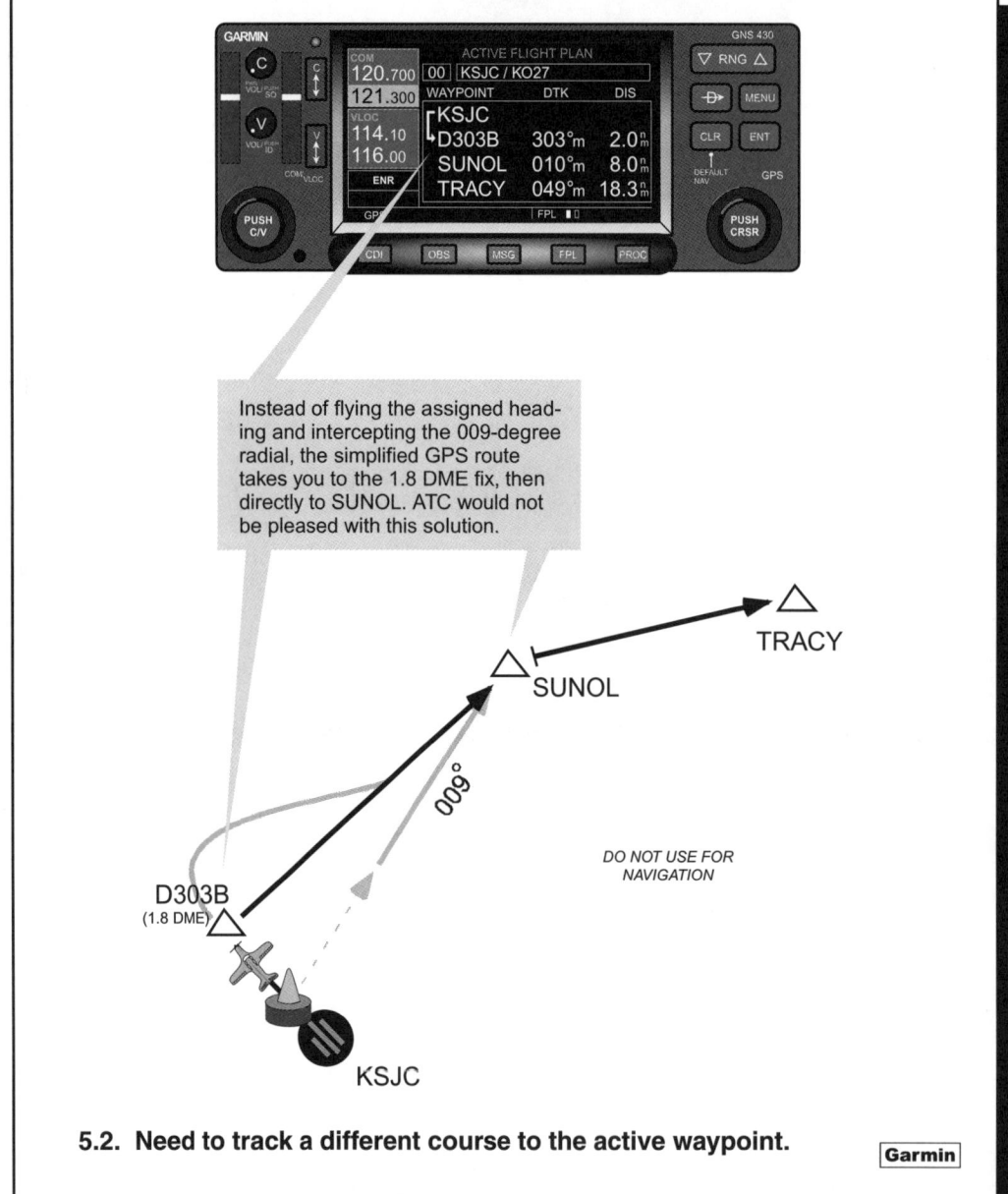

5.2. Need to track a different course to the active waypoint.

# 5 / Advanced Maneuvers

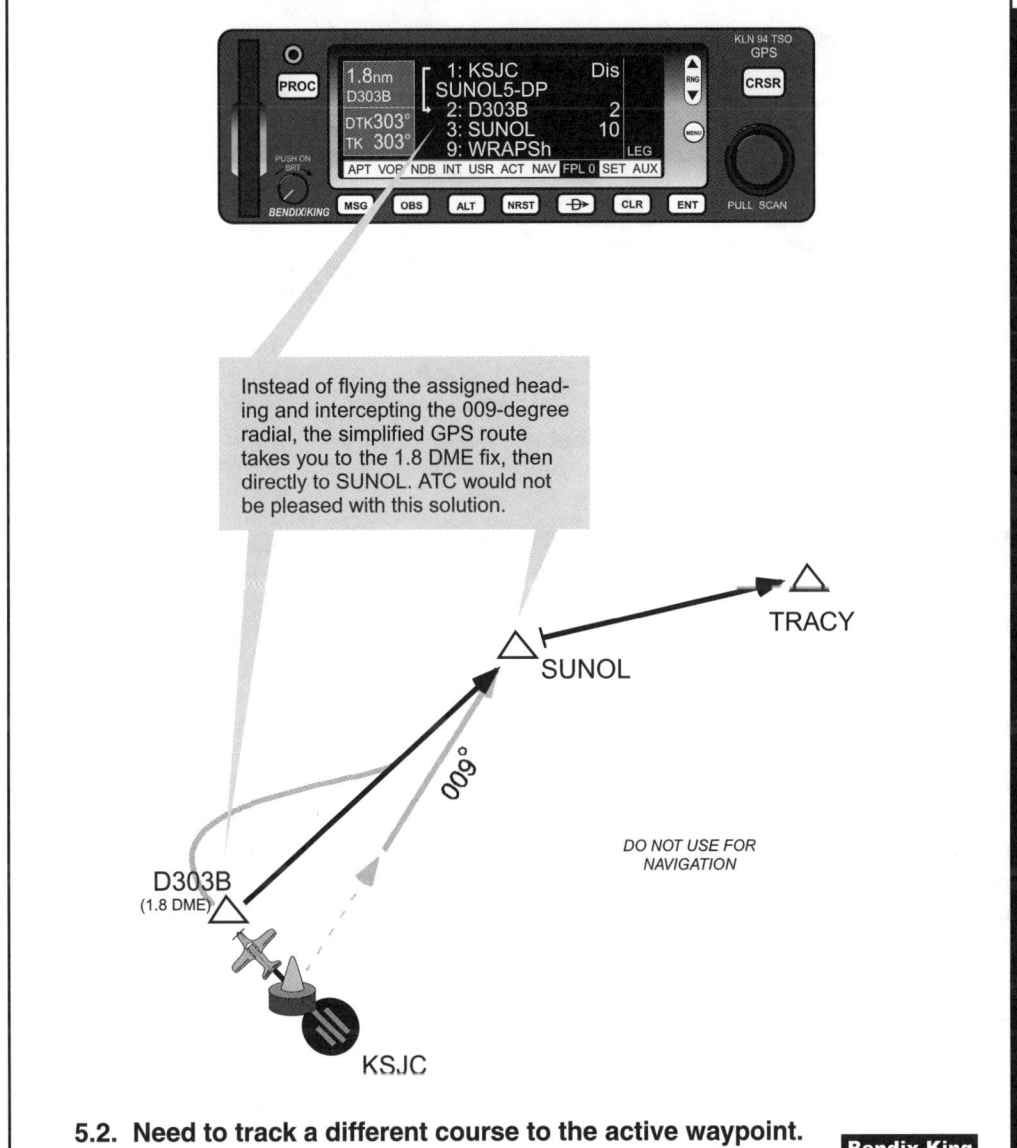

5.2. Need to track a different course to the active waypoint.

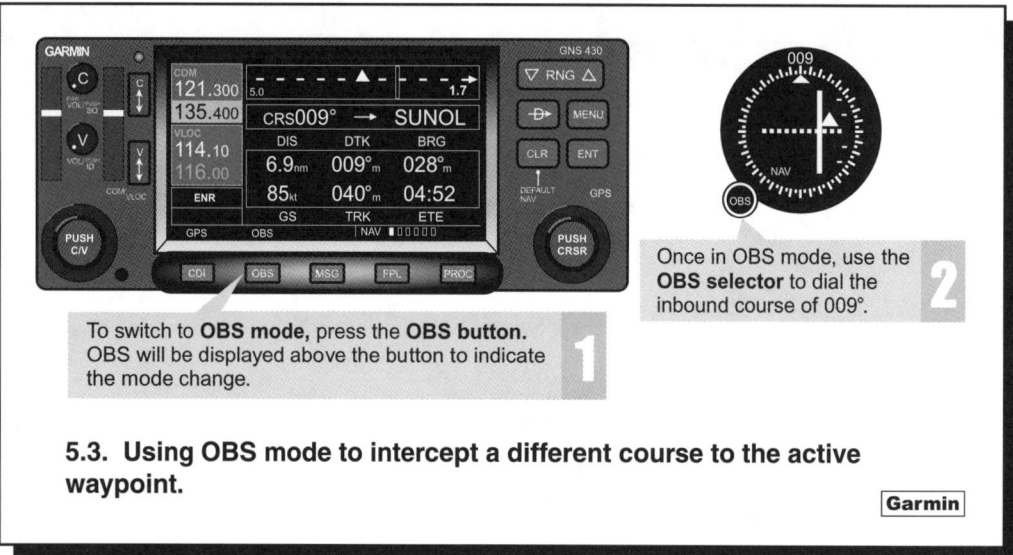

**5.3. Using OBS mode to intercept a different course to the active waypoint.**

### Engaging OBS Mode

Departing the San Jose airport, things go well until you reach the 1.8 DME fix (the D303B waypoint). Your disagreement with the route in the computer happens once you cross D303B. Here's the easy way to fix this problem. At D303B, turn to the assigned heading of 040 degrees and engage the nonsequencing (OBS) mode, as illustrated in Figure 5-3.

When you select OBS mode, you tell the GPS computer that you want to ignore the course to the active waypoint chosen by the computer and enter a different course of your choosing. OBS mode lets you specify whatever course you want to follow to get to Sunol. If you dial 009, the GPS computer will guide you along a 009-degree course into Sunol and disregard the direct course from D303B.

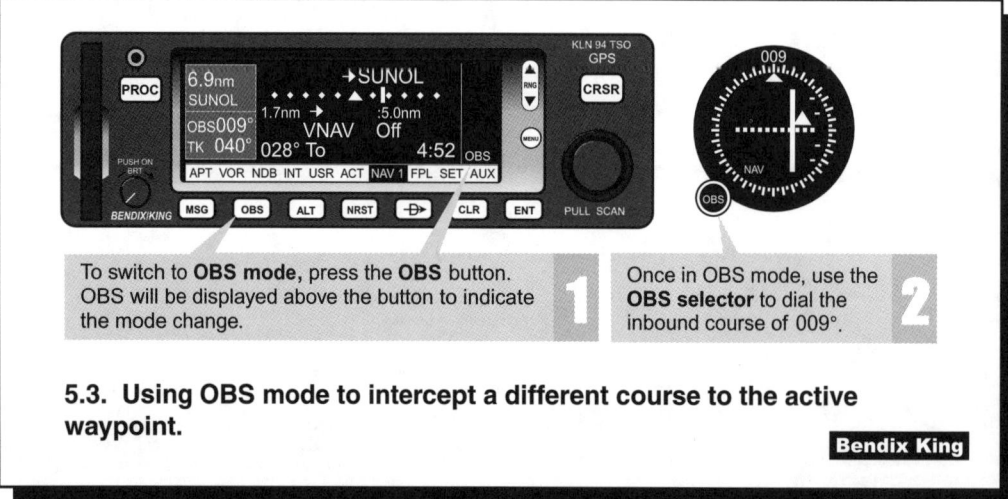

**5.3. Using OBS mode to intercept a different course to the active waypoint.**

# 5 / Advanced Maneuvers

Once you engage OBS mode and twist 009 using the OBS knob on your navigation indicator, the GPS computer provides you with guidance along the 009-degree course to the active waypoint. The navigation indicator in Figure 5-3 shows your position with respect to the 009-degree course. The navigation indicator shows that you are to the west of course. A 040-degree heading would seem to provide an ideal intercept heading, and this is precisely what the procedure requires. After proceeding on a 040-degree heading, the needle will center as you cross the 009-degree radial. You can then turn to a heading of 009 degrees and make progress toward Sunol intersection, the active waypoint.

Now that you are back on your route, you can switch the computer back into the sequencing mode prior to reaching Sunol intersection. Once back in the sequencing mode, the GPS computer will automatically sequence you to the next waypoint in the route after you reach Sunol. You will then be flying in the simple "connect-the-waypoints" sequencing mode as before.

It is useful to review what you have done. The GPS computer had planned for you to fly to D303B and then directly to Sunol intersection. You agreed that you should go to Sunol intersection but were not pleased with the inbound course. Therefore, you switched to OBS mode and dialed in a different course (009 degrees). You then intercepted and followed that course to Sunol.

Calling out the following two questions makes a great procedure for working your way through a situation in which you are asked to depart and rejoin your programmed route using OBS mode.

1. *Where am I going?* Point to the active waypoint on the navigation page and make sure it shows the waypoint that you wish to fly toward.
2. *How am I getting there?* Point out the desired track to the active waypoint on the navigation page. If it isn't the one you want, push the OBS button and dial the course that you do want.

Ask yourself these two questions all the time. It's a great way to stay on top of where the computer is taking you!

## SKILL 13: INTERCEPT AN INBOUND COURSE TO A DIFFERENT ACTIVE WAYPOINT

Now that you have learned how to dial in your own inbound courses, let's make things more interesting. Let's reconsider the Oakdale GPS approach you have been working with. Suppose prior to reaching the ECA VOR, Stockton Approach instructs you to fly a heading of 060 degrees, vector to the final approach course. Looking at the diagram of the approach in Figure 5-4, you can see that Stockton Approach is giving you a shortcut. You are to forget about going to ECA and make a beeline for the approach course somewhere between ECA and Moter intersection.

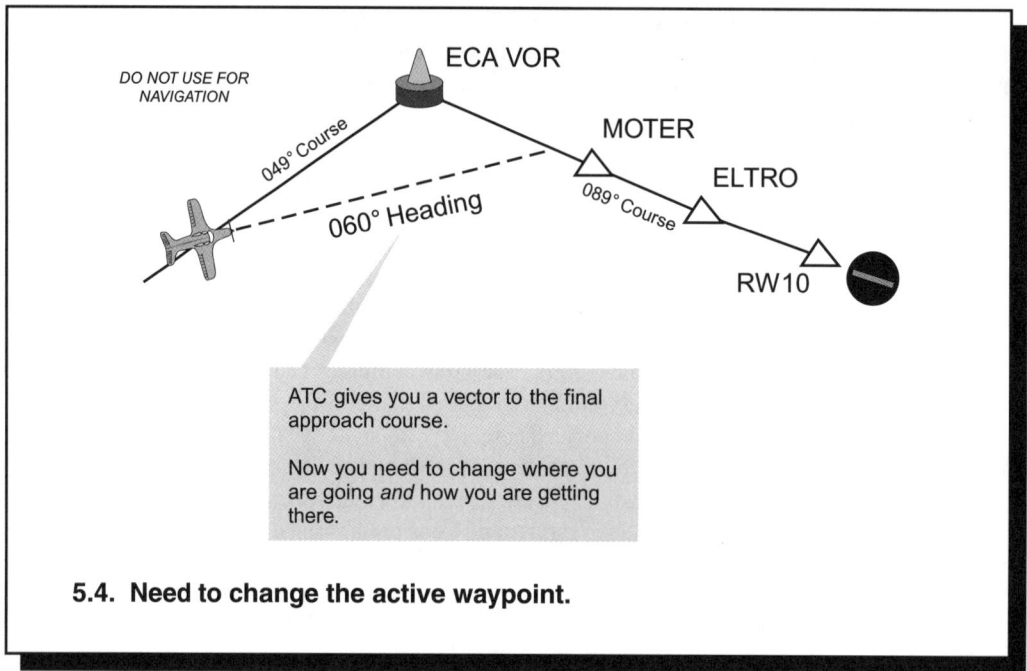

**5.4. Need to change the active waypoint.**

The first thing to realize is that you have just been asked to depart your programmed route and then get back on it again. OBS mode should immediately pop into your mind as the solution for this type of problem.

Let's begin by thinking about the two questions you need to always ask yourself when departing and rejoining your programmed route.

1. *Where am I going?* The next waypoint you want to go to is Eltro intersection. As you can see on the navigation page at the top of Figure 5-5, the active waypoint is currently ECA, but you do not want to go there, now or ever. You really want Eltro to be the active waypoint. How do you make this happen? How about your old friend the Direct-To function? If you go to the flight plan page, engage cursor mode, scroll to Eltro, press the Direct-To button, and verify it, Eltro now becomes the active waypoint (see Figure 5-5).

   Mission accomplished . . . sort of. If you look at the desired track to Eltro you can see that the GPS computer has defined a course from your present position directly to Eltro, a course of 070 degrees. This is not what you want to do. You want to continue on your heading of 060 degrees until you intercept the final approach course and then proceed to Eltro along the 089-degree published approach course. In other words, you have a good answer to the question of "Where am I going?" but you're not happy with your answer to . . .

2. *How am I going to get there?* To solve this problem, you simply do what you just learned how to do: switch to OBS mode and twist in the final approach course of 089 degrees. When you do this, you can see you have positive course guidance to the final approach course showing on the navigation indicator, as shown in Figure 5-6.

The answer to the question "How am I going to get there?" is now to follow the inbound approach course of 089 degrees to Eltro. You will maintain your assigned heading of 060 degrees until the needle centers and then turn inbound and reengage the sequencing mode.

# 5 / Advanced Maneuvers

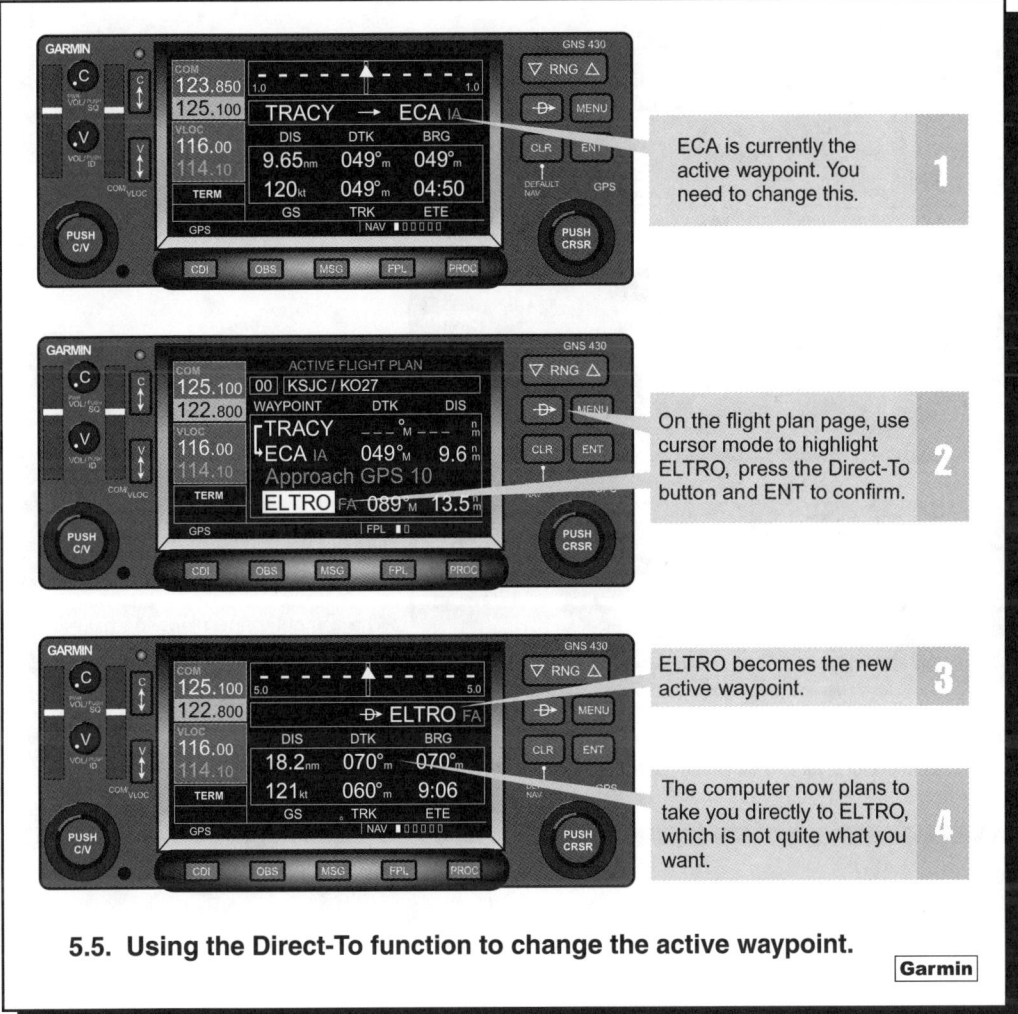

5.5. Using the Direct-To function to change the active waypoint.

**82** COCKPIT AUTOMATION for General Aviators and Future Airline Pilots

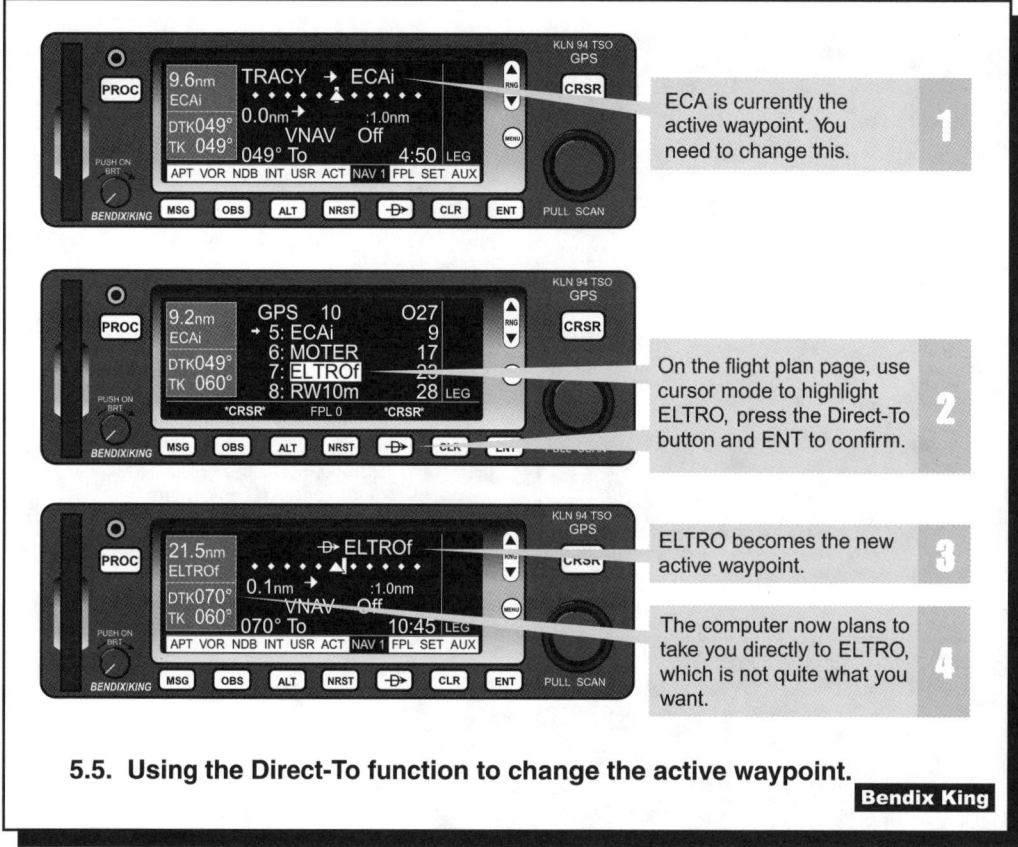

5.5. Using the Direct-To function to change the active waypoint.

5 / **Advanced Maneuvers** 83

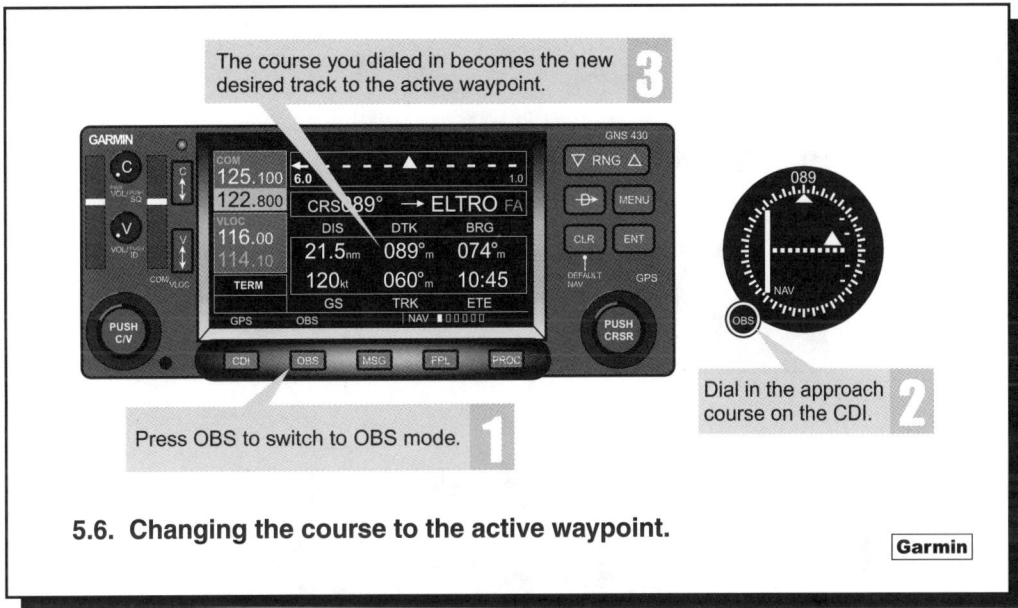

**5.6. Changing the course to the active waypoint.**

**5.6. Changing the course to the active waypoint.**

### The Vector-to-Final Feature

In an effort to avoid some of the potential confusion surrounding the task of switching between the sequencing and nonsequencing modes, many GPS computers offer an automated solution to the problem of flying an approach in which you receive vectors to the final approach course.

Let's revisit the scenario in which you are on your way to ECA when approach control gives you a vector to join the final approach course. You can use the vector-to-final feature in this situation following the procedure illustrated in Figure 5-7.

Press the PROC button and choose this option.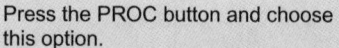

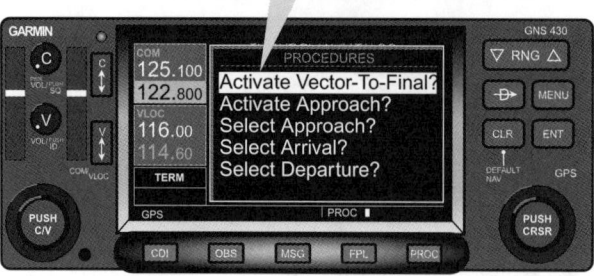

ELTRO is now the active waypoint, and the final approach course is set as the desired track. You will remain in Sequence mode throughout the approach. There is no need for mode switching.

Dial in the approach course on the CDI.

**5.7. The vector-to-final feature.**

# 5 / Advanced Maneuvers

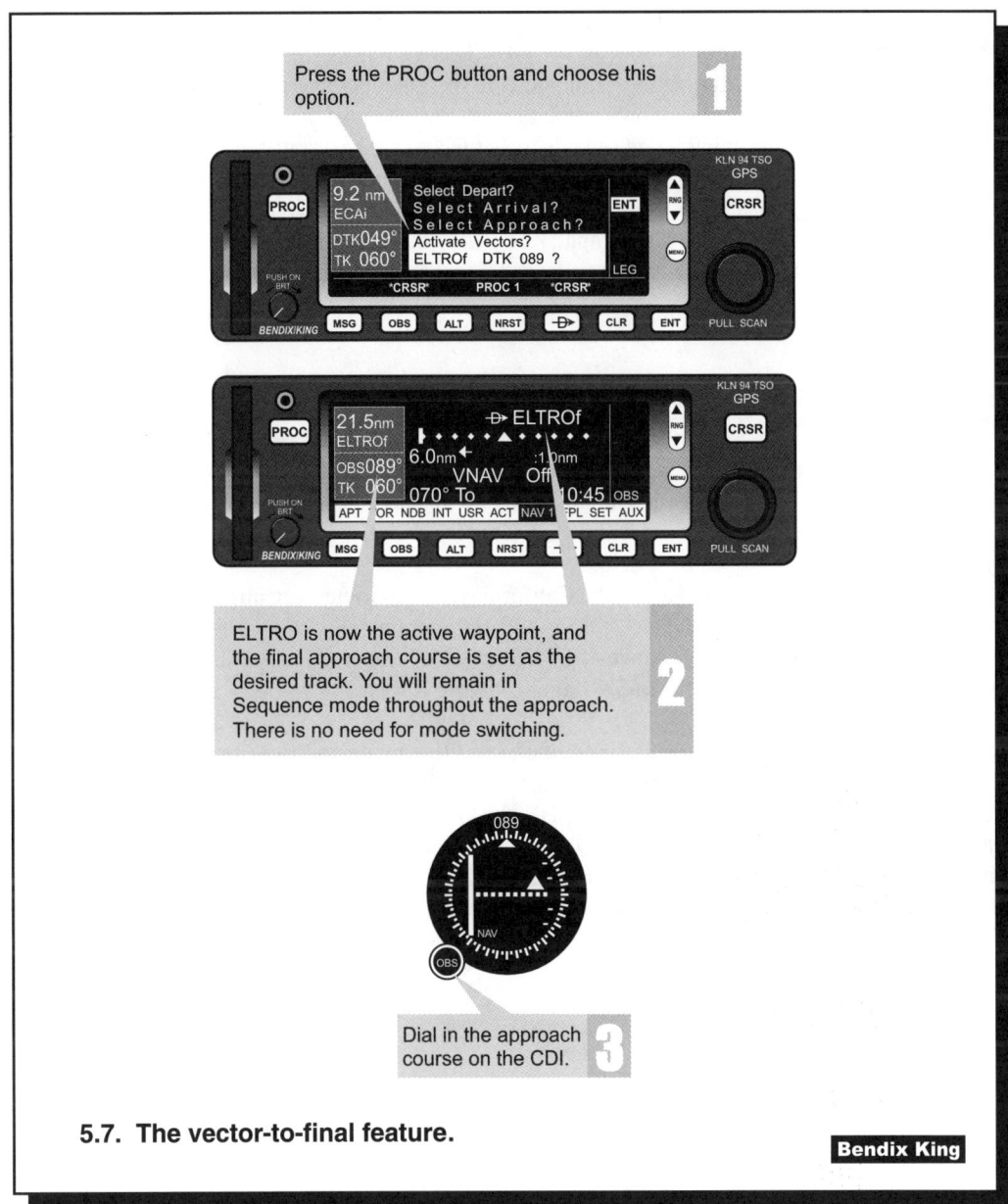

5.7. The vector-to-final feature.

When set to use the vector-to-final feature, the GPS computer automatically carries out the two familiar steps for intercepting an approach course: (1) set the FAF as the active waypoint and (2) dial in the final approach course as the desired track to the active waypoint. The one difference is that you remain in the sequencing mode at all times, and there is no need for mode switching.

The vector-to-final feature makes your job easy in those cases in which you don't mind eliminating the waypoints in your route prior to the FAF. There are many other cases in which you may rather leave some waypoints remaining in the route. In this case, perhaps it is best to manually choose your own active waypoint, dial your own inbound courses, and make the mode switches. In any case, it is best to master both techniques and use whichever technique best suits the situation at hand.

When using the vector-to-final feature, keep in mind that just because the computer is automatically performing steps that you would normally have to do yourself, you should still remain aware of what is going on at all times. No computer comes with a guarantee that it will work correctly every time. Ask yourself the same two questions about where you are going and how you plan to get there.

## SKILL 14: HOLDS

If OBS mode does not sequence after passing the active waypoint, it keeps giving you guidance to the same waypoint. It would seem that guidance like this might cause you to fly around in circles, crossing the same waypoint over and over again. This is precisely what you must do when asked to fly a hold.

Suppose you are southbound along the California coast en route to an airport near Los Angeles, and you are about to reach the Salinas (SNS) VOR as diagrammed in Figure 5-8. Prior to reaching SNS, ATC instructs you to hold northwest of SNS on V25. What you really want to do here is to fly to SNS, turn 180 degrees, and fly to SNS again. Figure 5-8 shows you how easy the OBS mode makes this job for you.

Prior to reaching SNS, put the GPS computer in OBS mode and twist in the inbound course. Once you reach SNS, turn 180 degrees, continue for 1 minute, and then turn back to track the needle inbound to SNS. Each time you fly over SNS, the GPS computer will not sequence onto the next waypoint in the route because you are in OBS mode. When you finally get released from the hold, simply put the computer back in the sequencing mode on your way inbound to the hold fix. This time, when you reach SNS, the GPS computer will sequence SNS through and the next waypoint in the route will become the active waypoint.

5 / Advanced Maneuvers

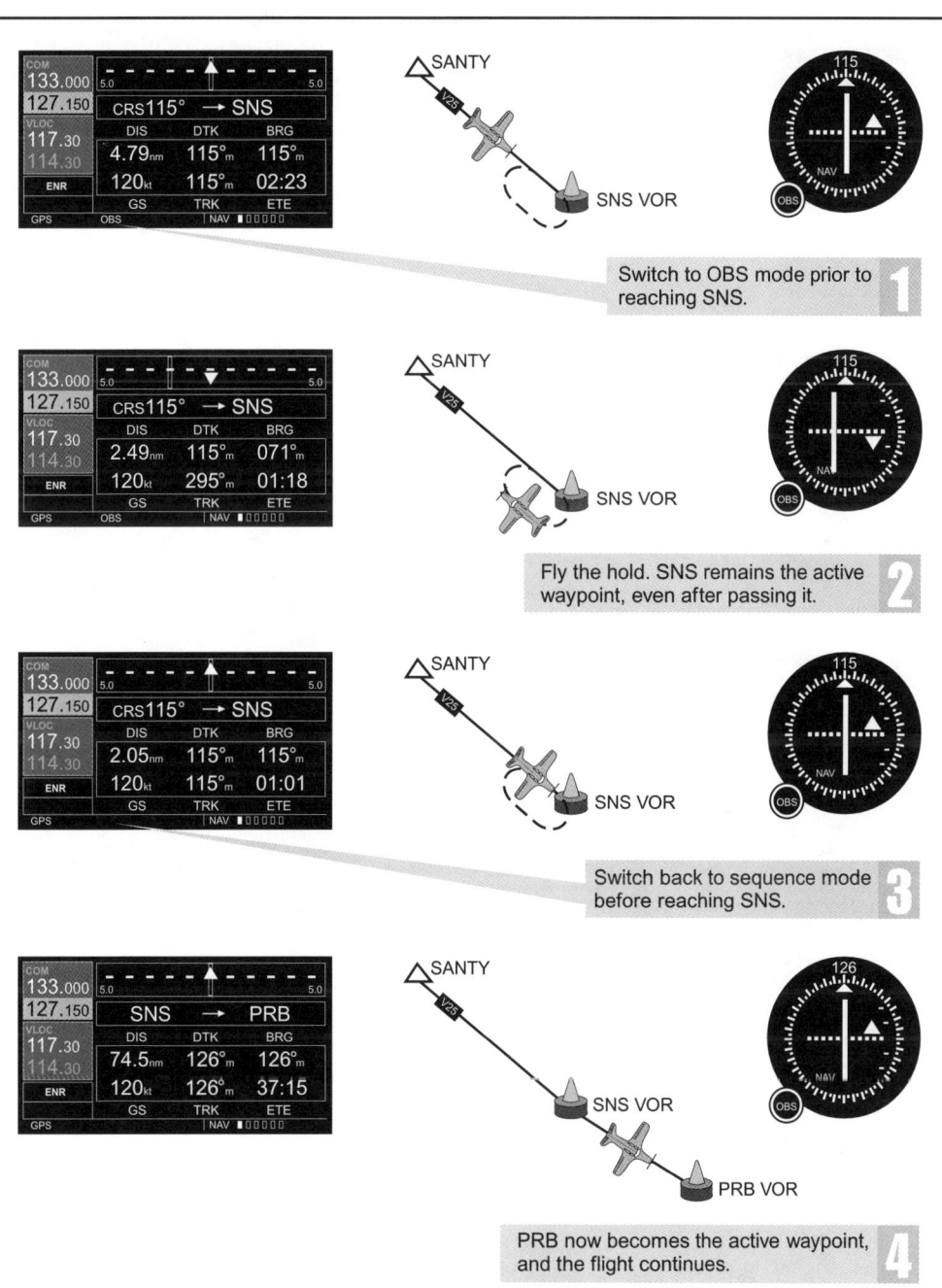

5.8. Flying a holding pattern with the nonsequencing mode.

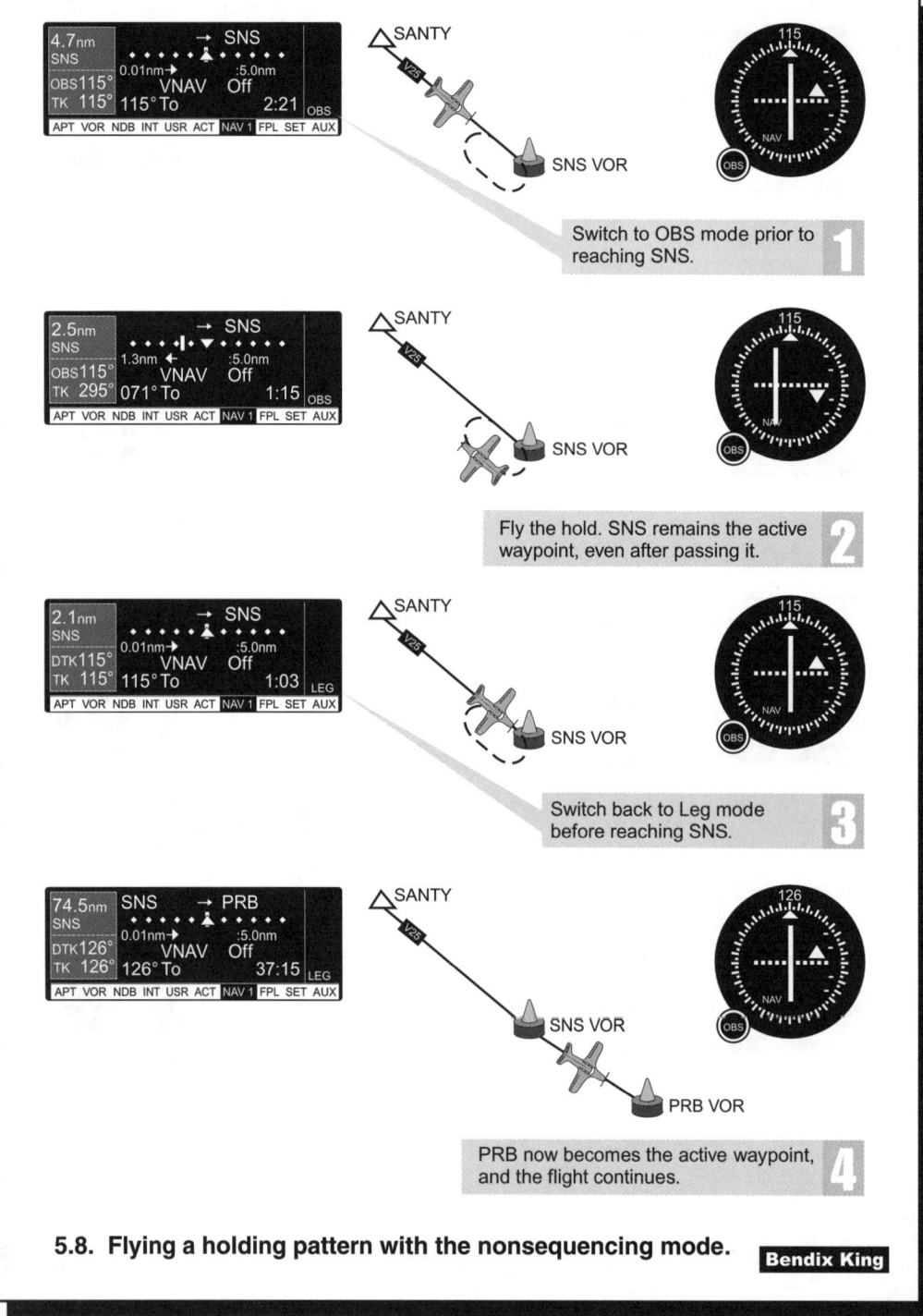

5.8. Flying a holding pattern with the nonsequencing mode. Bendix King

## SKILL 15: PROCEDURE TURNS

Another situation in which it is useful to turn off waypoint sequencing is the procedure turn. Let's choose the trickiest possible example of a procedure turn to illustrate the idea. Consider the approach into Lodi, California (1O3) diagrammed in Figure 5-9. This approach uses the familiar racetrack pattern as a procedure turn.

The waypoint called Quads is not only the initial approach fix (IAF), not only the final approach fix, but also the missed approach hold fix! Looking at the procedure in Figure 5-9, if you continue to Quads intersection in the sequencing mode, it would seem that the computer would recognize that you have passed Quads and immediately sequence you to the missed approach point. If you want to fly the procedure turn, you are going to have to get out of the sequencing mode before reaching Quads and prevent this from happening. The Garmin and the Bendix/King computers take two different approaches to solving this problem.

Illustrated in Figure 5-10, the Garmin computer *automatically* switches you out of the sequencing mode when you cross the initial approach fix. This gives you an opportunity to fly around the racetrack once before turning back to the final approach fix and continuing on to the missed approach point. After flying the procedure turn and on your way back to the final approach fix, the Garmin computer *automatically* switches you back to the sequencing mode and guides you to the missed approach point. Do not make the mistake of pushing the OBS button before or after reaching the IAF.

Illustrated in Figure 5-10, the Bendix/King computer takes a different approach to solving this problem. The Bendix/King computer leaves you in charge of the mode switching. Prior to reaching the initial approach fix, the Bendix/King computer will suggest the idea of putting the computer in the nonsequencing mode. This suggestion takes the form of a message ("If Required Select OBS"). This message reminds you to put the computer in OBS mode prior to reaching the

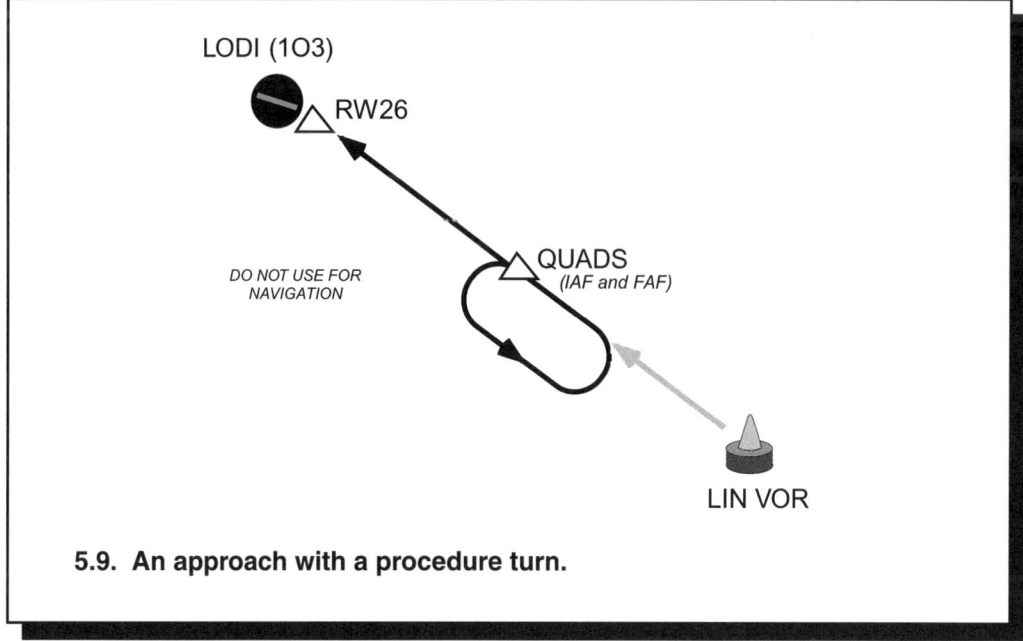

5.9. An approach with a procedure turn.

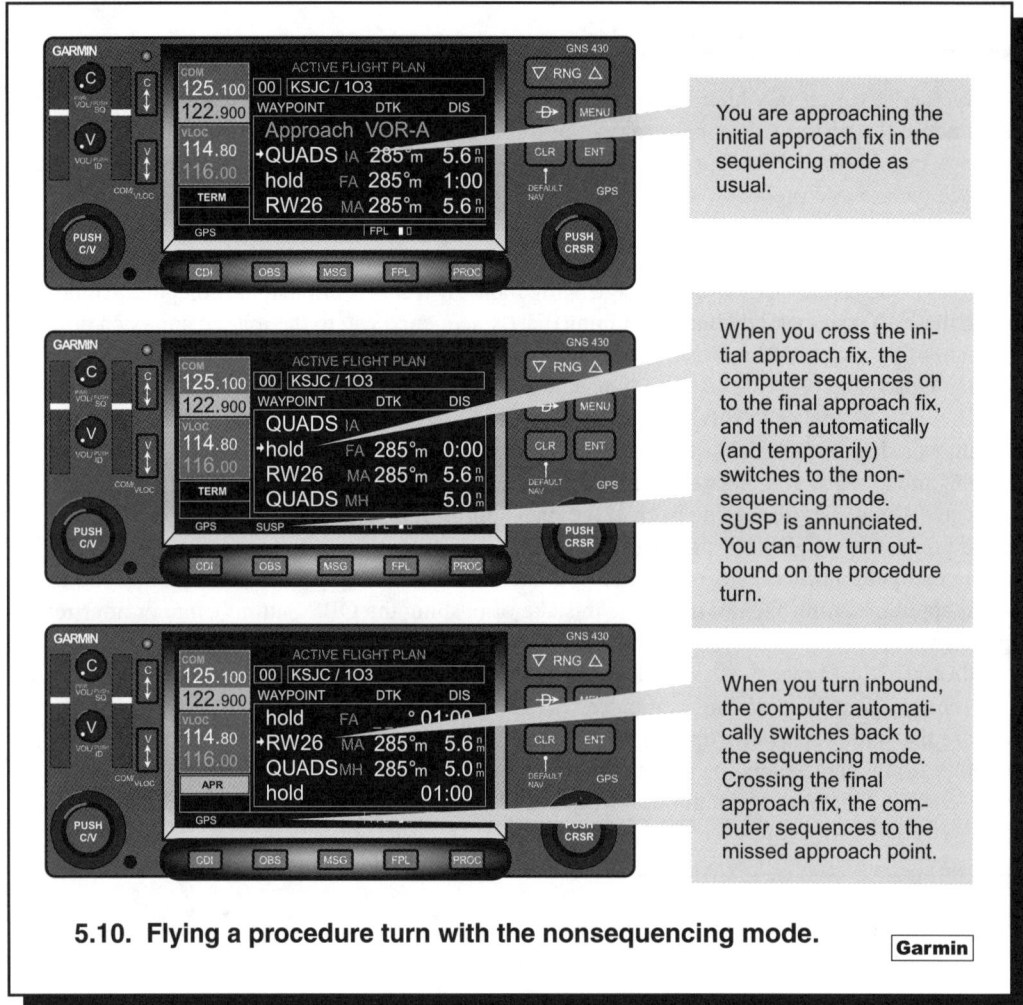

5.10. Flying a procedure turn with the nonsequencing mode.

initial approach fix. Switching to OBS mode will allow you to fly your procedure turn and then turn back inbound again to put the computer back into Leg mode and continue your approach. Do not make the mistake of neglecting to push the OBS button before you reach the initial approach fix. This missed step will prompt the computer to sequence immediately to the missed approach point and skip the procedure turn.

Note how the two computers take different approaches to the problem of reducing the possibility of making an error. One computer leaves you with the possibility of making what we call an **error of omission** (neglecting to take a needed step). The other computer presents the possibility of making an **error of commission** (taking a step when one wasn't needed). Many years of human factors research have taught us that humans are prone to committing both of these errors. Consequently, neither computer has the ultimate solution to getting it right every time. The answer for now: practice makes (almost) perfect!

The same technique for flying the racetrack-shaped procedure turn can be used for flying the other kinds of procedure turns. Buttonhooks, teardrops, or racetracks: The computer doesn't care much what you do in between the mode switches. When engaged in the nonsequencing mode, final

# 5 / Advanced Maneuvers

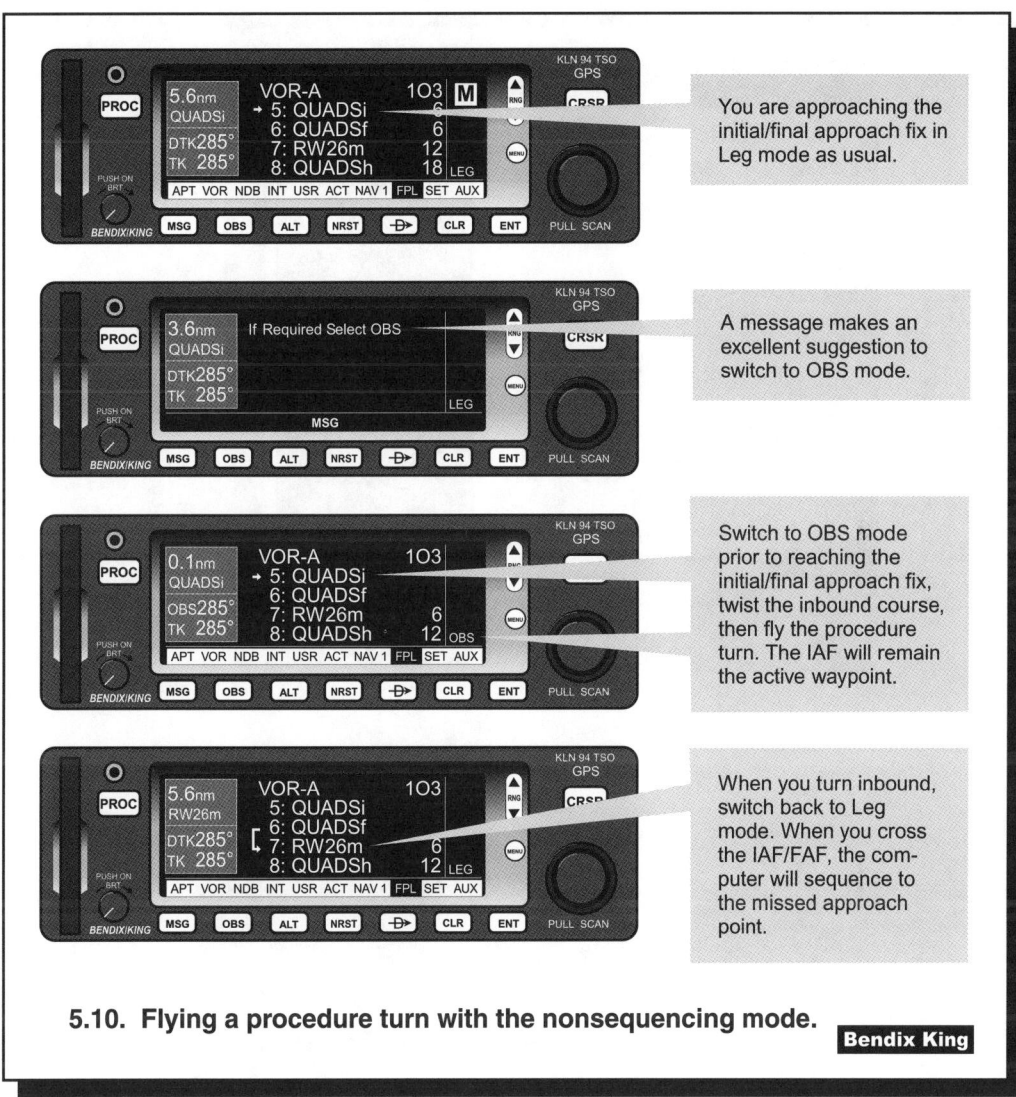

**5.10. Flying a procedure turn with the nonsequencing mode.** Bendix King

approach fix will remain the active waypoint, and the computer will continue to give you guidance along the final approach course.

## SKILL 16: MISSED APPROACHES

Let's talk about yet another situation in which it is useful to turn off waypoint sequencing. Suppose you reach the missed approach point on the GPS Runway 28R approach into Monterey, California, and decide you must execute the missed approach. The missed approach instructions are "Climb to 1,900 feet, then climbing right turn to 6,000 feet, direct SNS VOR and hold." As you approach the missed approach point, the flight plan page lists the missed approach point followed by the remaining waypoints in the missed approach procedure. Figure 5-11 illustrates the missed approach procedure and the associated waypoints on the flight plan page.

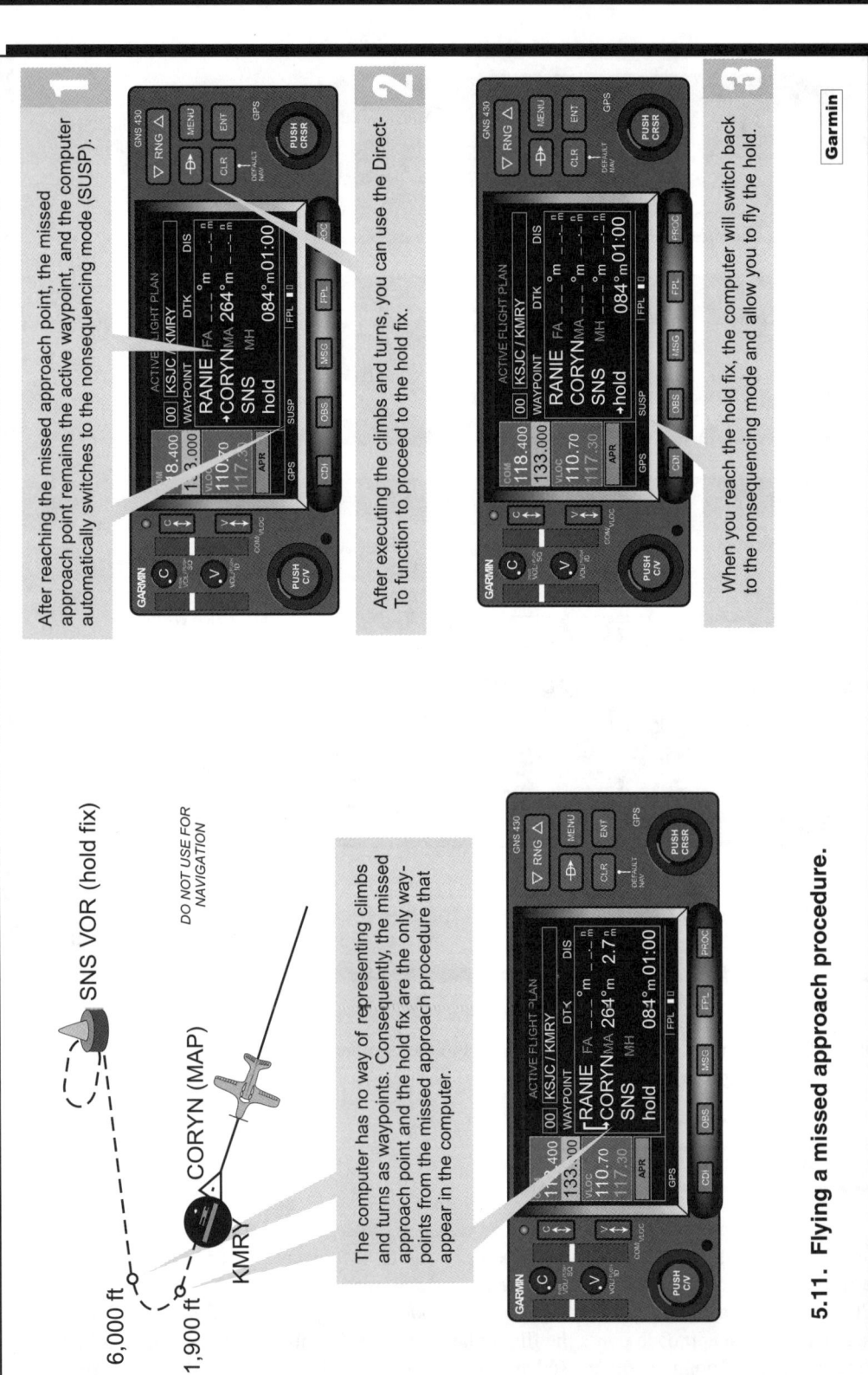

5.11. Flying a missed approach procedure.

# 5 / Advanced Maneuvers

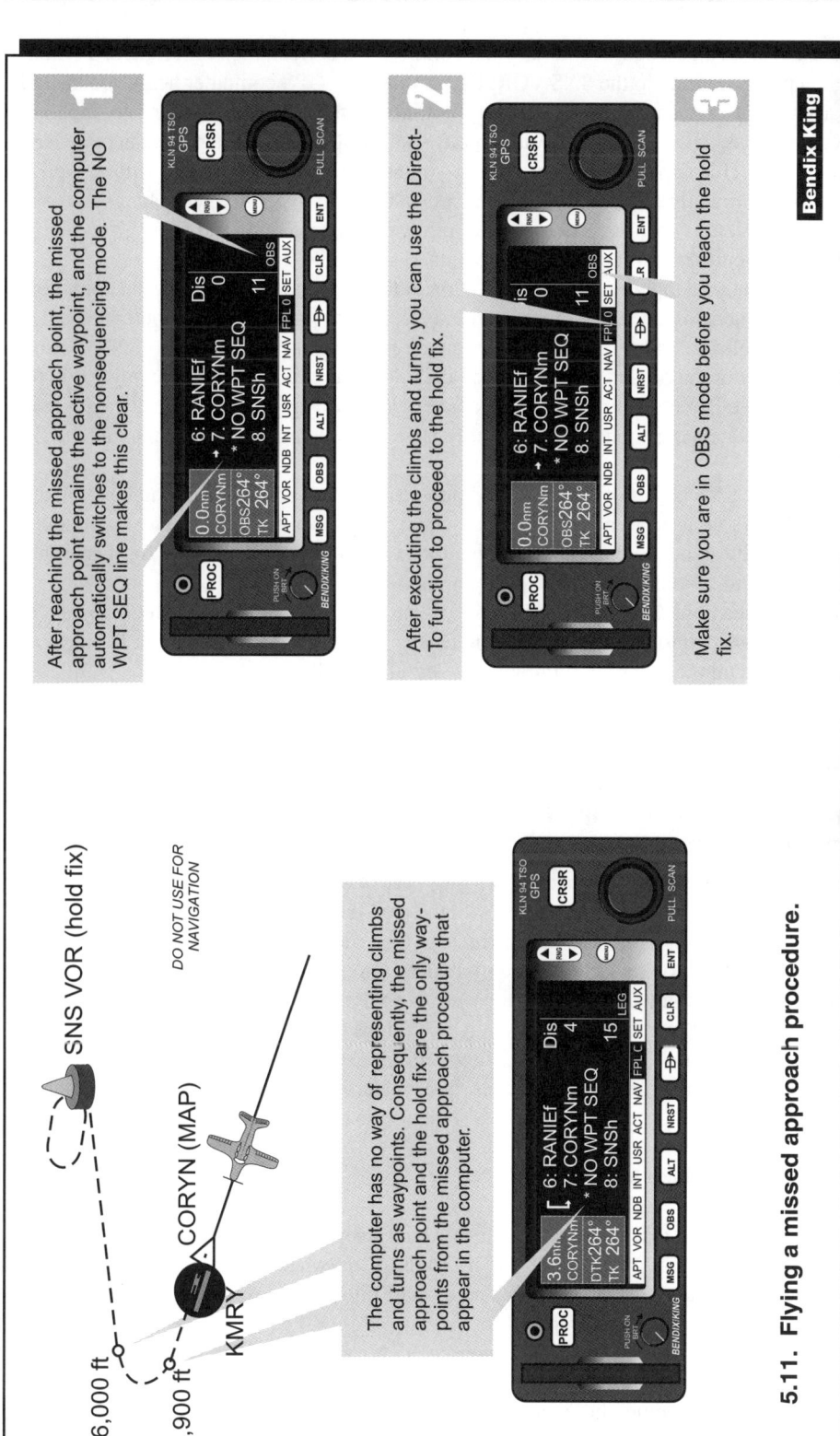

5.11. Flying a missed approach procedure.

The missed approach procedure requires you to climb to 1,900 feet, then turn right and climb to 6,000 feet, then proceed direct to the SNS VOR. Recall that your GPS computer helps you navigate by playing a game of connect-the-waypoints. By definition, these waypoints are geographically fixed locations in the sky. But where will your aircraft reach 1,900 feet on the missed approach procedure at Monterey? This depends on what aircraft you are flying. A Cessna 172 will be well out over the water by the time it reaches 1,900 feet. A Cessna Citation may only get as far as the end of the runway. The problem is that there is no way to represent these climbs and turns given the way your GPS computer represents routes using waypoints.

Both GPS computers adopt the following solution to the problem: the computers automatically turn off waypoint sequencing once you reach the missed approach point. Each computer then allows you to execute your climbs and turns, and when you are ready, you can proceed to SNS. For convenience, a waypoint for the hold at SNS is already included in the route. After reaching 6,000 feet, you can simply use the Direct-To function to make the SNS hold waypoint the active waypoint and then reengage the sequencing mode. You will now have sequencing mode guidance to the SNS VOR.

Since the hold at SNS is part of the published missed approach procedure, it can be carried out using the same technique used to perform the procedure turn described above. The Garmin computer will automatically switch to the nonsequencing mode when you reach the hold fix. The Bendix/King will advise you to switch to the nonsequencing mode.

Once in the missed approach hold, you'll have yet another task: deciding where to go next and programming these plans into the GPS computer. This is a high-workload situation: having to fly the airplane while planning and programming a new leg of your flight.

## SKILL 17: EXPLAIN THE IMPORTANCE OF CONTINUALLY MONITORING THE GPS COMPUTER

Experience has shown that the most challenging part of using cockpit automation is making sure that the plan you have in your mind agrees with the plan that is programmed into the computer, and that this plan agrees with what ATC has instructed you to do. It should come as no surprise that the possibility for mix-ups looms large. The only real way to protect yourself against this type of confusion is to constantly compare your plan for what to do next with what the GPS computer is set up to do next. Let's look at two common ways in which things can go wrong when the communication between human pilot and computer breaks down. In these two examples you will see that both pilot and computer can sometimes make mistakes, and an alert pilot can prevent them from leading to disaster.

### Computers Sometimes Make Mistakes

In this example, we will discuss just one case in which a sophisticated computer can sometimes do something unexpected. Our purpose is not to try to make an exhaustive list of all the mistakes a computer can make, but rather make the point that a smart pilot knows that these things can sometimes happen and is always alert and prepared for them.

Consider the case of the GPS Runway 28R approach into Modesto, California. Suppose you are en route to Cazli intersection and decide to fly a heading to join the airway between Cazli and Awoni intersection as diagrammed in Figure 5-12.

# 5 / Advanced Maneuvers

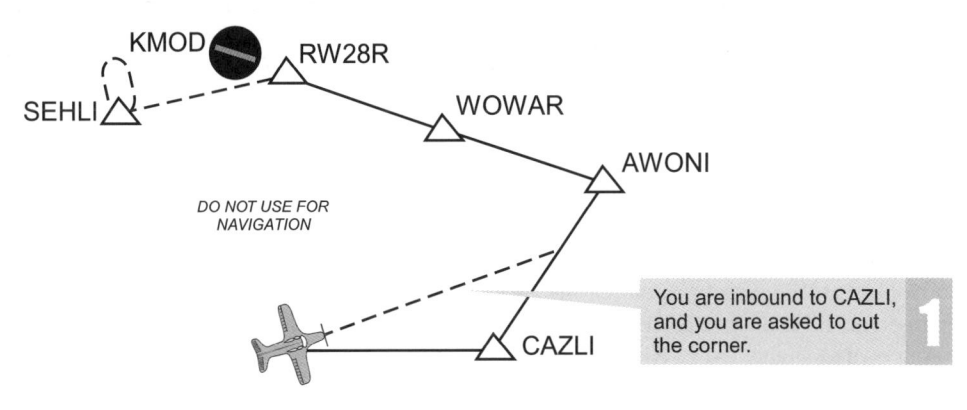

You are inbound to CAZLI, and you are asked to cut the corner. **1**

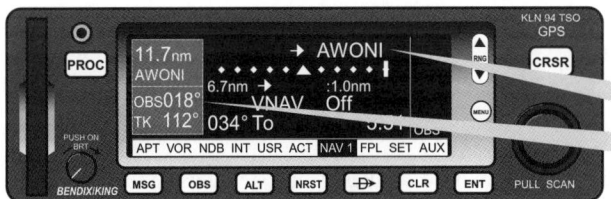

You make AWONI the active waypoint, switch to OBS mode, and twist in the new course. **2**

You switch back to sequence mode early, and the computer sequences to the missed approach point. **3**

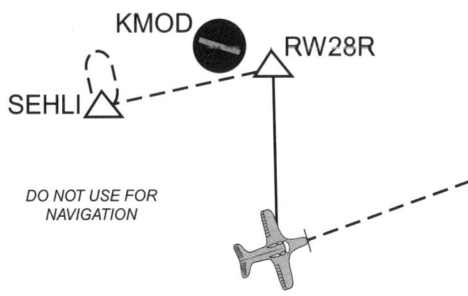

**5.12. Computers sometimes make mistakes.**

Bendix King

You switch to OBS mode and twist the inbound course to Awoni. Guidance to the course between Cazli and Awoni is now showing on your navigation indicator. Realizing that you are now simply waiting for the needle to center so you can turn inbound to Awoni, you decide to switch back to the sequencing mode so as not to forget to do it when you intercept and turn inbound to Awoni. You know that if you forget to reengage the sequencing mode before Awoni, it won't sequence the waypoint and make Wowar the active waypoint. So you reengage the sequencing mode now and look at the navigation page. You see that RW28R is now the active waypoint and your guidance shows a direct course to RW28R. Your entire approach has disappeared, and you have no approach guidance whatsoever. What happened? Waypoint sequencing happened. The waypoint sequencing function looked at the current position of the aircraft and the route and tried to determine what you wanted to do next. From your present position, the closest waypoint to you is the missed approach point, as shown in Figure 5-12. When you reengaged the sequencing mode, waypoint sequencing assumed that you had already passed Cazli, Awoni, and Wowar and were intending to go to the missed approach point. You seemed to be a little south of the approach course, not to mention heading in the wrong direction, but that can be fixed with a simple turn to the left.

One moral of this story is: Follow along with what your computer is doing at all times because you may not always like what it is doing.

### Pilots Sometimes Make Mistakes

In this example, we look at a case in which the pilot makes a mistake, and the GPS computer is now programmed to do something undesirable.

Suppose you are on the same approach into Modesto and get the same vector to intercept the approach course (see Figure 5-13).

You use OBS mode to intercept the course inbound to Awoni, but this time you forget to put the computer back into the sequencing mode. Since you are still in OBS mode, you will not get waypoint sequencing. That is, once you reach Awoni, the GPS computer will not give you guidance to turn toward the airport. Rather, you will continue straight ahead. What if there was rising terrain in this direction? As you can see, this simple error could lead to disaster.

### To Err Is Human, to Be Error-Tolerant Is Divine

Study after study has shown that not maintaining awareness of the intentions of an automated cockpit system is the single most important obstacle to becoming a proficient pilot in a modern cockpit. There are several explanations for why this may be so. First, the problem of monitoring and configuring an automated system like a GPS computer is something completely new to the pilot who has never flown with automation. Pilots are now being asked to use something with which they have no prior experience. Second, radios, VORs, DMEs, and ADFs are relatively straightforward devices. Automated systems that show subtle differences on their displays are often configured to do radically different things. In the example above, the difference between completing the approach and continuing into the terrain was the difference between a few annunciations on your instrument panel. Indeed, human factors experts have argued that lack of feedback is one of the critical flaws in the design of current automated systems.

How can you stay out of trouble? First, start with a good understanding of the sequencing and nonsequencing navigation modes. Know how they work, their limitations, and in what situations they are best used and in what situations their use should be avoided. Second, develop the attitude that mistakes are going to be made by both pilot and computer and develop good awareness habits that make you prepared to detect and fix them when they happen. The verbalization procedure of

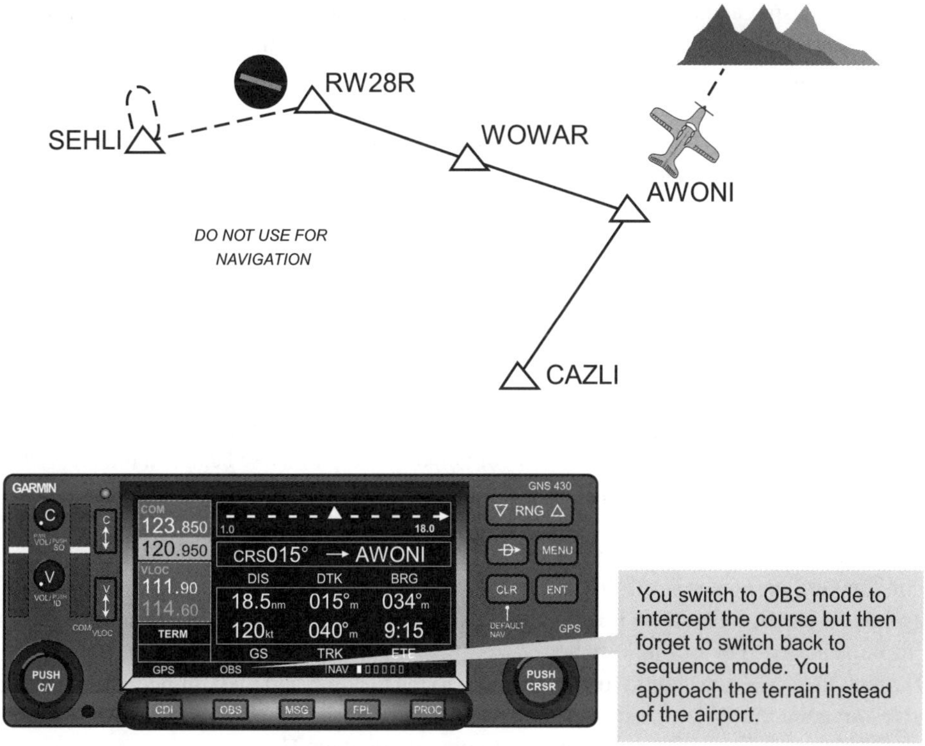

## 5 / Advanced Maneuvers

**5.13. Pilots sometimes make mistakes.**

(Figure callout: You switch to OBS mode to intercept the course but then forget to switch back to sequence mode. You approach the terrain instead of the airport.)

asking where you are going and how you are going to get there is a helpful one. Checklists are a great way to help ensure that progress monitoring and mode checks happen at critical times during challenging maneuvers such as vectored intercepts. Make use of all available navigation equipment in the aircraft to ensure position awareness in the event that the GPS computer does something confusing or unexpected. Be prepared for standalone GPS approaches in which guidance to your missed approach point and beyond may be particularly difficult when unexpected events occur with the GPS computer. Take the time to prepare for human and computer errors just as you would train for situations such as engine failures: It's no secret that humans and computers break down more often than airplane engines. Be prepared and safe, not surprised and sorry.

## PRACTICE SESSION (AIRPLANE AND/OR SIMULATOR)

The skills you learned in this chapter are the hardest to master and the most important to know how to do well. This practice session gives you the opportunity to practice using the nonsequencing mode of the GPS computer. You can do this practice session using a simulator or an airplane. If you were thinking of using an airplane for at least one of these practice sessions, this session is a good choice. Work through the flight scenarios presented in the chapter using the simulator first, then try your skills in the airplane.

Your flight should start with a departure procedure that gives you the opportunity to fly a vector, depart the programmed route, use the nonsequencing mode, and rejoin the programmed route. If possible, the flight should be planned to an airport at which at least one GPS approach can be flown, each approach commencing with vectors to the final approach course rather than overflying the published IAF. This practice session also gives you the opportunity to review route programming procedures and position awareness.

Particular attention should be paid to the problem of mode awareness during this practice session. It has been demonstrated that the development of mode awareness skills presents an ongoing difficulty for many students. You are encouraged to make regular call-outs of which modes are currently engaged and any immediate future plans for switching to different modes. For example, when proceeding on a vector to intercept the approach course, you are encouraged to call out that you are in OBS mode and plan to switch to the sequencing mode once the course has been intercepted. Prior to the FAF, try to verbally anticipate the arrival at the 2-NM point and that you have readied the aircraft for the final approach segment.

Attention should also be given to the concept of having some form of positive course guidance set up on the GPS computer at all times. You should keep in mind that at no time does ATC issue heading vectors alone; a clearance limit is always given, and this should be set up on the GPS as the target to be achieved. For example, when issued the clearance "Fly heading 090, vectors to join the 229 radial to ECA," you should turn the aircraft to the heading and then immediately make ECA the active waypoint and twist the desired course of 049. You should always have an answer to the questions "Where am I going?" and "How am I going to get there?"

Prior to departing, review the concepts behind the nonsequencing mode. Use the itinerary for the flight and verbally walk through the specific steps that will be followed for each vectored intercept maneuver.

### SKILL 12: Intercept an Inbound Course to the Active Waypoint

1. If no published departure procedure for the airport of origin exists, a mock departure procedure can be made up. The important point is that somewhere along the departure procedure, you should fly a vector to intercept and then rejoin the programmed route. The vector can take you back onto the programmed course prior to reaching the active waypoint or after the active waypoint. If the vector is to an intercept beyond the active waypoint, you will be required to change the active waypoint in addition to choosing the nonsequencing mode and dialing a new course.

### SKILL 13: Intercept an Inbound Course to a Different Active Waypoint

1. Perform two approaches using vectors to the final approach course. In each case, choose a new active waypoint in addition to selecting the nonsequencing mode and dialing the desired course.

### SKILL 14: Holds

1. Choose any waypoint along your route and give yourself a hold clearance. Try a direct entry first, then make up a more challenging hold.

## SKILL 15: Procedure Turns

1. Choose an approach that has a procedure turn and practice flying the full approach.

## SKILL 16: Missed Approaches

1. Execute the missed approach and hold for at least one of the approaches.
2. Program a route to a different destination while you are in the hold.

## SKILL 17: Explain the Importance of Continually Monitoring the GPS Computer

1. As you make your way from waypoint to waypoint, each time you take some action using the GPS computer (e.g., switch between modes, twist new courses), ask yourself the question "What would be the consequences of getting this step wrong and not noticing it?"

### How Did You Do? (Common Errors)

How did you do on this flight? Did you remember to call out the GPS approach modes on these approaches?

The most common errors for this flight are as follows:

1. Forgetting to get back into the sequencing mode after you have intercepted the inbound course.
2. Not choosing the correct active waypoint.
3. Not twisting-in the right inbound course to the active waypoint (many students mistakenly twist-in the assigned heading).
4. Beginning the missed approach prior to the missed approach point and confusing the waypoint sequencing function.
5. Not positively setting a course to the hold waypoint.
6. Not setting the computer to the nonsequencing mode prior to reaching the hold waypoint.
7. Not twisting the inbound hold course after reaching the hold waypoint.
8. Programming a new destination and approach while flying a missed approach hold.

 **DVD DEMONSTRATION DISC**

Skills 12, 13, 14, 15, 16, and 7 are demonstrated using the Garmin GNS 430 on the accompanying DVD disc.

# CHAPTER 6

# Flying with an Autopilot

*In this chapter, you will become familiar with a second kind of computer in the cockpit. An autopilot supports you in steering the airplane along the flight route you have programmed into your GPS computer. In exchange for this service, you will have to gain a basic understanding of how the autopilot works and serve as a supervisor while the autopilot does its work in flight.*

Thus far in the book you have learned to use your GPS computer to support you while performing a job known as **navigation.** You learned to use your GPS computer to plan a flight route, monitor your progress as you made your way along your flight route, and make changes to your route when directed to do so by ATC. Until now, the job of steering the airplane along the route you have planned has been left to you. In this chapter, we introduce another kind of cockpit automation system that helps you steer the airplane along the planned route. You will see how this system, when combined with the GPS computer, results in a very automated form of flight that places you, the pilot, in somewhat of a managerial role as the automation works its magic.

## THE AUTOPILOT

An **autopilot** is a system that is capable of helping you steer the airplane along the flight route you have planned. Perhaps the best way to explain how an autopilot works is in the context of a flight. Suppose you are making your way along the route to Oakdale airport that you have planned and programmed into your GPS computer.

Looking at the GPS computer shown in Figure 6-1, you can see that you are progressing toward Tracy intersection. The altimeter shows that you are flying at 5,500 feet. Let's assume that you have leveled off and have properly trimmed the airplane for level flight. The GPS computer shows that you are operating in the sequencing mode, meaning that the GPS computer will provide navigation services to you between each of the waypoints along your route. Also, you have pressed the CDI

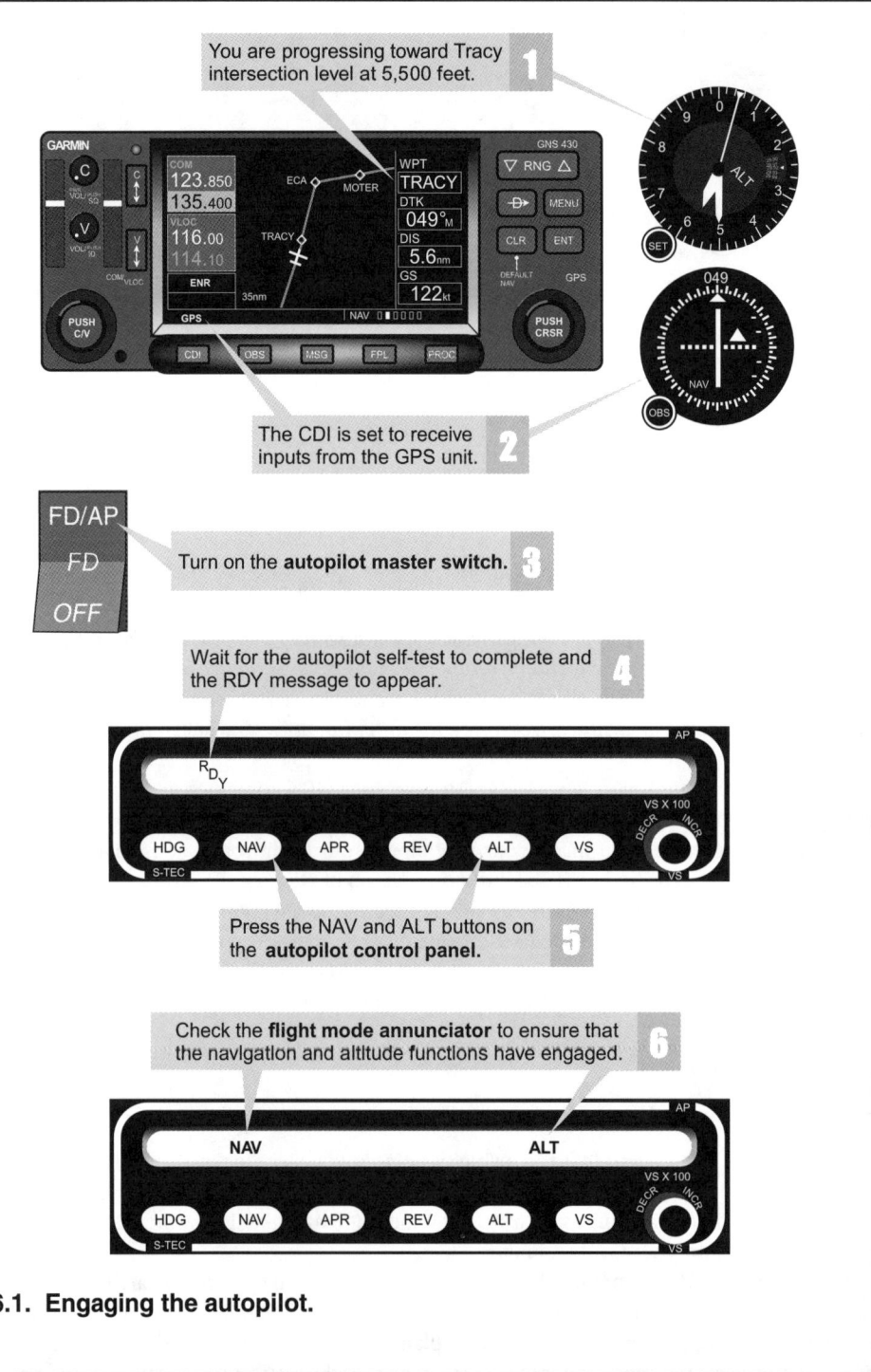

6.1. Engaging the autopilot.

# 6 / Flying with an Autopilot

button on the GPS computer to ensure that your position with respect to the desired track to each waypoint is presented on the airplane's navigation indicator.

Let's put the autopilot to work now and get some assistance in steering the airplane along the programmed route. Turn on the **autopilot master switch** shown in Figure 6-1. For now, flip the autopilot master switch all the way up to the position marked FD/AP. With the autopilot master switch turned on, the autopilot will go through a series of self-tests and display a message when it is ready to be used.

Turning your attention now to the **autopilot control panel** (also shown in Figure 6-1), you can see from the buttons on the panel that the autopilot offers several different functions. Being an experienced pilot, you can probably guess what most of these buttons mean, but for now, let's concentrate on two of them. Press the button marked NAV, then the button marked ALT, and let go of the control yoke. You will find that the airplane is now flying itself, almost completely automatically.

How is this impressive feat being performed? It's quite simple, really. The autopilot has been given two important **targets** to achieve. One target comes from the GPS computer and is shown on the CDI. This target is the desired track of 049 degrees to Tracy intersection. The CDI needle tells you whether you are on target or not. The autopilot uses the CDI to determine how well it is doing in achieving the target. If the CDI needle is centered, everything is going according to plan. If the CDI needle strays to one side or the other, the autopilot must then determine what control yoke manipulations must be carried out in order to return the airplane to the desired track to Tracy intersection. In the case that the airplane has strayed from the desired track, the autopilot will figure out what roll manipulations are required to turn the airplane to a reasonable intercept heading, recapture the desired track, and then turn the airplane back on course. This is the function performed by the autopilot's **navigation function,** the one you engaged when you pressed the NAV button on the autopilot control panel.

The second target that the autopilot is working toward is maintaining the airplane at the established cruising altitude. When you pressed the ALT button on the autopilot control panel at 5,500 feet, the autopilot assumed the task of maintaining the airplane at that target altitude. This is the function performed by the autopilot's **altitude function.**

With the navigation and altitude functions engaged, the autopilot will steer the airplane along the desired track to the active waypoint that is presented on the CDI and at the altitude at which you engaged the altitude function. As the airplane reaches Tracy intersection, the autopilot will automatically turn the airplane to follow the desired track to the new active waypoint in the route.

## Staying in the Loop

As you can now see, using the autopilot and the GPS computer together results in a fairly automated kind of flight. If the GPS computer and autopilot are automatically deciding where to go and also controlling the airplane to take you there, what is the pilot's role in all of this? The answer is that you have now been placed in the role of supervisor or manager. Rather than viewing this situation as a good opportunity to work in an afternoon nap, you must actively monitor the progress of the airplane while the automation performs its duties. You must constantly ask yourself the following questions: What active waypoint am I working toward? What course am I tracking to get there? Who or what is steering the airplane? What does the automation plan do once I reach the active waypoint?

For the pilot who is now left with the task of monitoring a collection of computers in the cockpit, one of the most useful features of the GPS computer is that it provides plenty of feedback about what it is currently doing and what it plans to do next. In previous chapters, you learned how to use the many pages of the GPS computer to keep track of the activities of the GPS computer. Let's now look at the problem of keeping track of what the autopilot is doing.

### The Flight Mode Annunciator

To help keep you in touch with what's going on, every autopilot has some form of **flight mode annunciator** (**FMA**). The flight mode annunciator shows which autopilot functions are currently engaged. The autopilot control panel shown in Figure 6-1 displays the name of any autopilot function that is currently engaged, just above the button used to engage the function. Some airplanes additionally feature a separate FMA that can be located near the airplane's primary flight instruments to make it a part of the pilot's normal scan. It is particularly important to rely on the flight mode annunciator as your sole means of determining which autopilot functions are currently engaged. You will experience many cases in which you push a button in hopes of engaging an autopilot function, only to discover (or not!) that the function did not engage. You will hear airline instructors say over and over again: "Check your FMA!" It's the only true way of determining what your airplane is doing.

## THE FLIGHT DIRECTOR

Some autopilot systems are combined with another system that helps keep the pilot informed about the activities of the autopilot. A **flight director** displays the roll and pitch commands being generated by the autopilot and presents them to the pilot. A flight director is shown in Figure 6-2.

Integrated with the familiar attitude indicator, the flight director adds **command bars** to the display. Command bars show the pitch and roll commands that the autopilot has decided that the airplane needs to carry out in order to achieve the targets it has been given. When the airplane symbol on the attitude indicator is lined up with the command bars, the airplane is following the pitch and roll manipulations that the autopilot is generating.

The flight director in Figure 6-2 is presented on a new kind of flight instrument. An **electronic flight instrument system** (**EFIS**) replaces the traditional primary flight instruments. EFIS integrates aircraft attitude, altitude, airspeed, vertical speed, heading, and turn rate into one or two combined displays. Note how the integration of the flight instruments changes the way that instruments are scanned.

### Flying with the Flight Director Only

If you think about it, the autopilot really performs two jobs when automatically steering the airplane. First, at each step of the way, the autopilot has to decide what control manipulations are required. For example, when reaching the active waypoint and turning to join and follow a new airway, the autopilot has to figure out that roll adjustments are needed to turn the airplane and that pitch must be adjusted slightly to compensate for the loss of lift in the turn. Second, once the autopilot has decided what control manipulations are needed, it must carry them out. In a sense, we can say that the autopilot has both brains and muscles. The brain figures out what needs to be done, and the muscles work the controls to make it happen.

Having a flight director presents you with an interesting alternative for how to fly the plane. It turns out that autopilots that also have flight directors allow you to configure the system in such a

# 6 / Flying with an Autopilot

**6.2.** A flight director presented on an electronic flight instrument system (EFIS) display.

way that the brain is turned on, but the muscles are turned off. That is to say that you can set the autopilot to figure out what control manipulations are required at each step of the way, have these needed roll and pitch manipulations presented on the flight director, but then leave it to you to actually work the control yoke to make the needed control inputs happen. In this situation, you are acting as the autopilot's muscles. Your job in this situation is to watch the flight director command bars and work the control yoke to keep the airplane symbol on the attitude indicator lined up with the command bars. This presents an interesting situation in which you are put in the role of following orders given by an autopilot. Aside from following the flight director commands, it is critical that you remain aware of the overall plan of the GPS computer and autopilot. You should never assume that the flight director commands represent a master plan that is free from errors.

## MANEUVERING WITH THE AUTOPILOT

So far you have learned how to use the autopilot to follow the route programmed in the GPS computer in level flight. But what about the other maneuvers you are required to do in the course of the typical flight? You will now learn to use the autopilot to perform climbs, descents, intercepts, vectors, and all kinds of instrument approaches.

### Climbs and Descents

Suppose you are en route to Oakdale airport when ATC instructs you to plan to cross ECA at 3,000 feet. You flip over to the vertical navigation page on the GPS computer and plan out your descent. You decide on a 500-foot-per-minute descent rate and see that you are to start your descent in about 5 minutes. The problem is that with the altitude function engaged, the autopilot will maintain the airplane at 5,500 feet, the altitude at which the function was originally engaged. How can you accomplish your 500-foot-per-minute descent to ECA?

The autopilot's **vertical speed function** allows you to perform constant-rate climbs and descents. The vertical speed function is illustrated in Figure 6-3.

As you reach the top-of-descent point planned by the GPS computer, simply reach up to the autopilot control panel, press the VS button, and dial $-500$ feet. The autopilot will then manipulate the pitch of the airplane to maintain a descent rate of 500 feet per minute.

Keep in mind that there are two other chores that need to be taken care of when performing any climb or descent. First, the power needs to be adjusted. Unless your airplane has an **autothrottle system,** you must adjust the power to an appropriate setting yourself. Second, the airplane may need to be retrimmed. When this is the case, your autopilot will provide you with **trim commands.** Figure 6-4 shows the autopilot control panel commanding an upward and downward adjustment of the pitch trim.

# 6 / Flying with an Autopilot

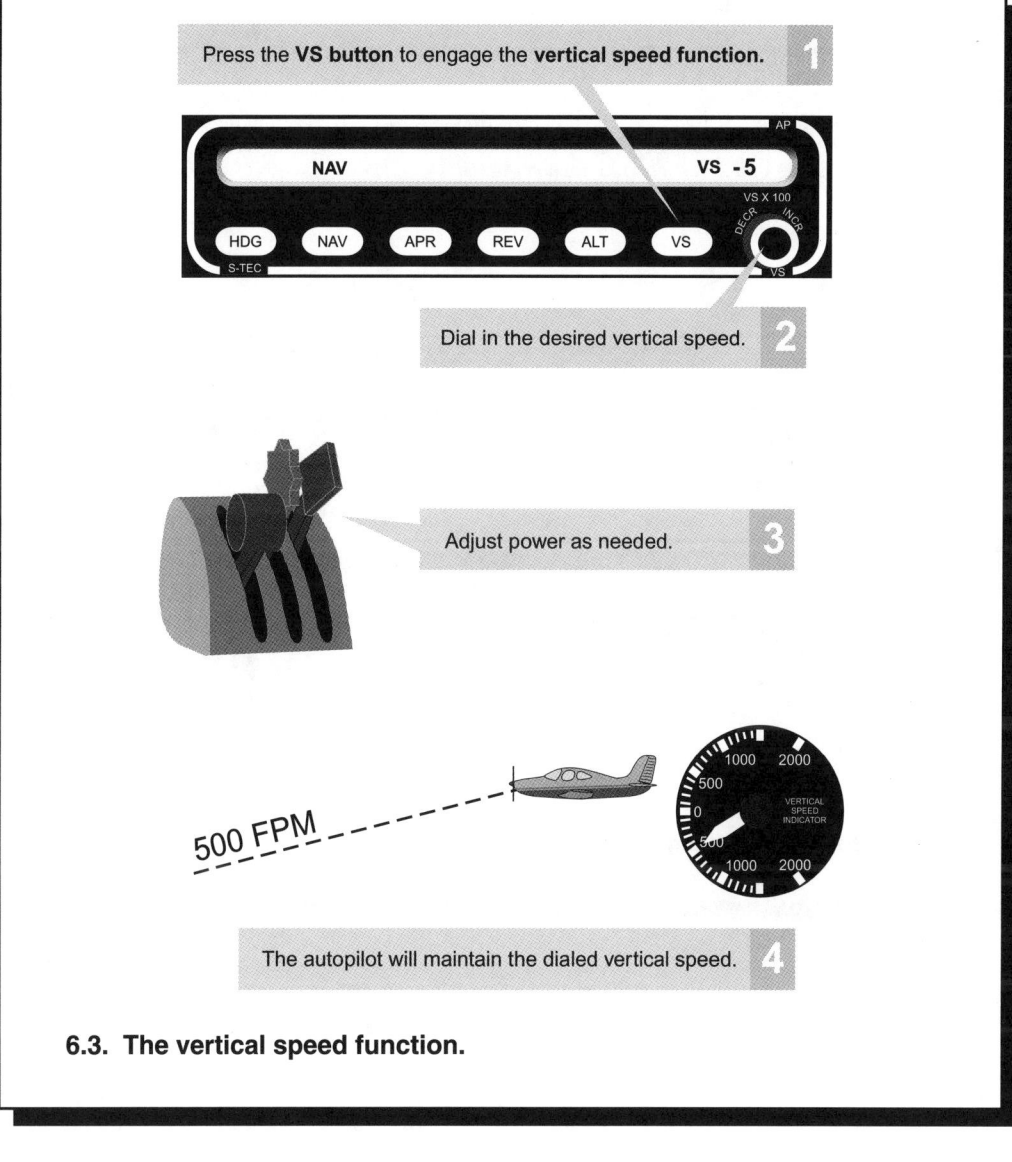

6.3. **The vertical speed function.**

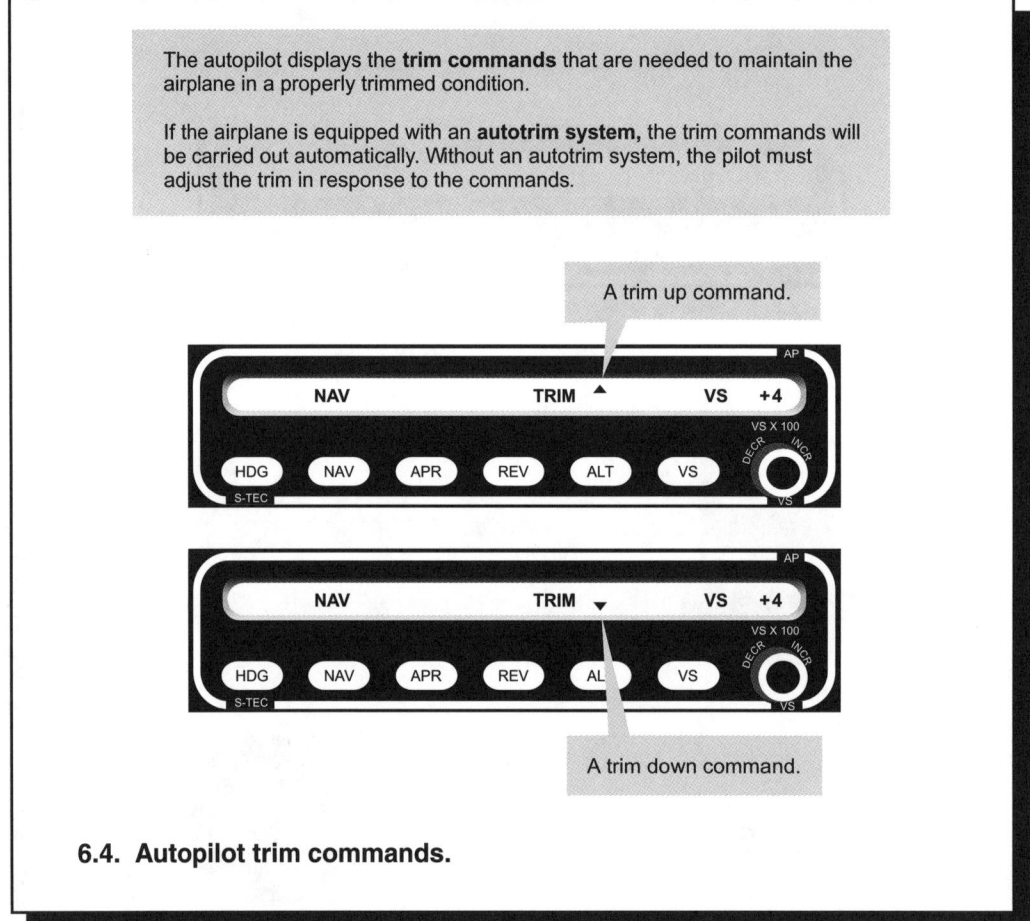

6.4. Autopilot trim commands.

In many airplanes, you must follow these commands and adjust the trim yourself. In airplanes equipped with an **autotrim** system, as its name suggests, the autopilot is capable of automatically making any needed adjustments to the pitch trim to maintain the airplane at the desired altitude and in a properly trimmed condition. When an autotrim system is installed, the autopilot still presents trim commands on the autopilot control panel. The only difference is that the trim commands are being carried out automatically.

Some autopilots can be combined with an **altitude selector/alerter.** An altitude selector/alerter allows you to dial an assigned altitude when using the vertical speed function. Just prior to reaching the dialed altitude, the altitude alerter will provide a visual and auditory alert that the airplane is approaching the desired altitude. Upon reaching the dialed altitude, the autopilot will perform an **altitude capture** maneuver and automatically level the airplane. Figure 6-5 illustrates the process of using the altitude selector/alerter to dial and capture an altitude with the vertical speed function.

# 6 / Flying with an Autopilot

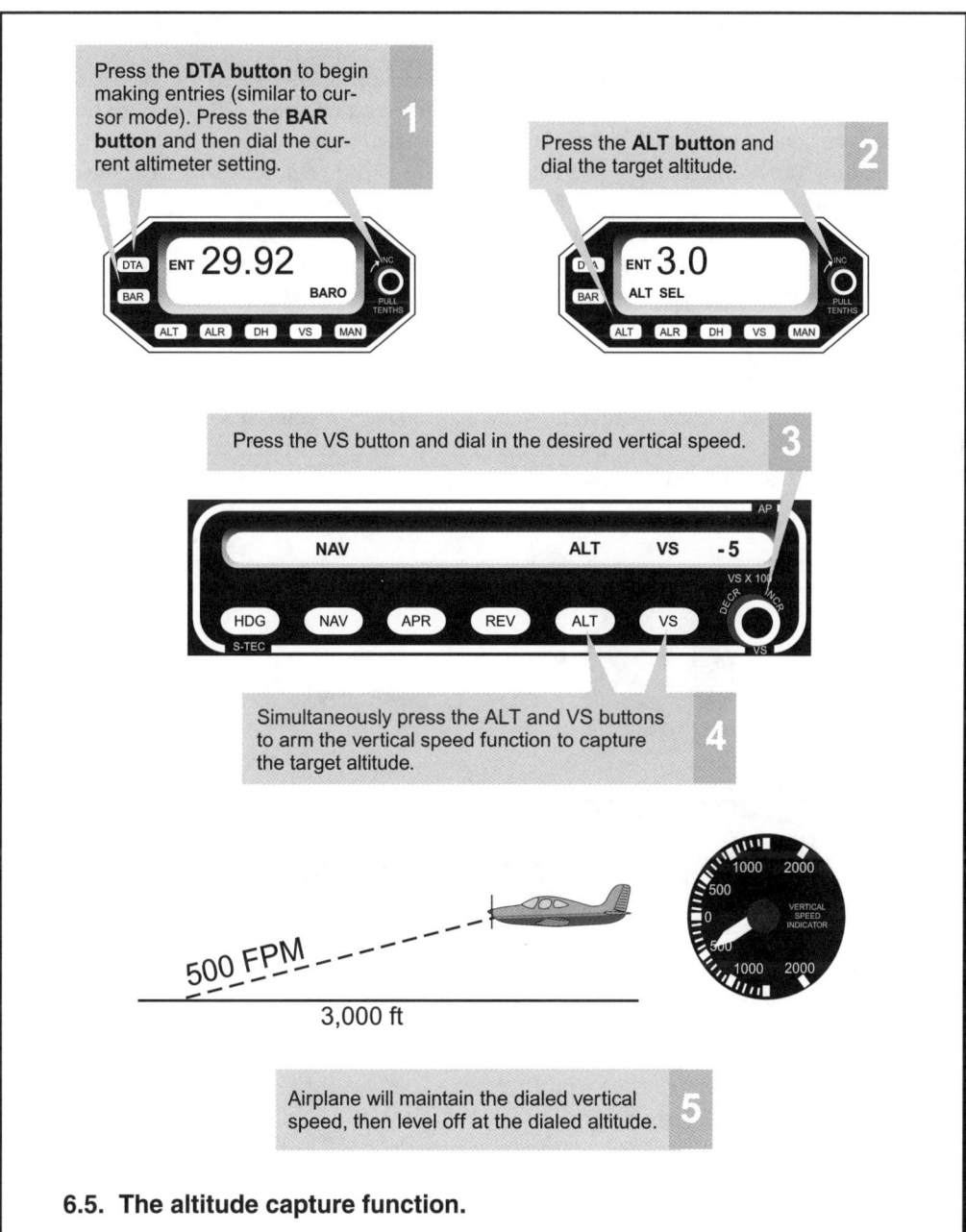

**6.5. The altitude capture function.**

### Flying an Assigned Heading

Now suppose you are coming up on ECA and descending to 3,000 feet. ATC calls and tells you to fly heading 060, vectors to the final approach course. You set up your GPS computer for the intercept using the nonsequencing (OBS) mode and now need to fly the assigned heading.

The **heading function** offered by the autopilot can be used to automatically steer the airplane along any heading that is dialed using the heading bug that appears on the heading indicator. If the airplane is not currently steering along the heading bug, or the heading bug is moved, the airplane will enter a maximum 25-degree banked turn until the desired heading is reached. The heading function is illustrated in Figure 6-6.

### Tracking a VOR Radial

Suppose you decide to leave behind the route programmed in your GPS computer and opt for good old-fashioned VOR navigation. It turns out that the autopilot's **navigation function** also allows you to track VOR radials in addition to the route programmed into your GPS computer. Before

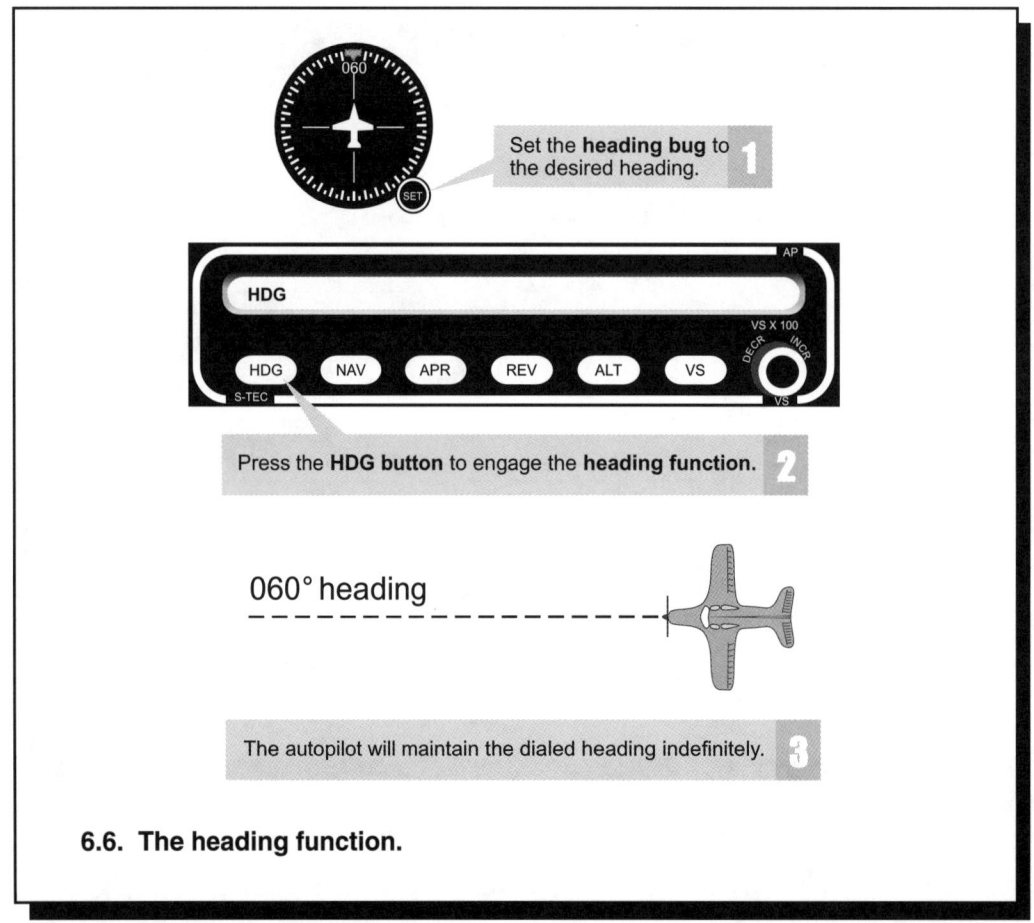

**6.6. The heading function.**

using the navigation function to track a VOR radial, you'll need to do two things that you haven't done yet in this book.

First, you will have to tune and identify the VOR station. You are probably already quite familiar with navigation radios. If you are using the Garmin computer, it contains a built-in communication and navigation radio receiver. You will likely have little trouble finding your way around this easy-to-use radio, and you can always consult the manufacturer's manual for more details.

Second, you will have to set the navigation indicator to display your position with respect to the VOR radial you have dialed. Throughout the book, you have had your navigation indicator set to display GPS course information. Now it's time to try the other setting.

Suppose you are arriving at the ECA VOR and are asked to track the ECA 089-degree radial outbound. You can comply with this request using the autopilot's navigation function, as illustrated in Figure 6-7.

Be sure to use the same caution when dialing VOR radials with the autopilot that you would when tracking VOR radials yourself. The autopilot will allow you to commit the same "reversal errors" that pilots are sometimes prone to committing themselves. For example, if you wish to track outbound on the 089-degree radial, you need to dial 089 into the CDI. If you accidentally enter the reciprocal, the autopilot will execute a 180-degree turn and track the 089-degree radial inbound, just as you have requested.

### Intercepting and Tracking a VOR Radial

Let's now make the radial intercept problem a little more difficult. Suppose that you have been vectored toward the Oakdale airport, off the airway that leads to the ECA VOR. You are then asked to intercept and then track the same 089-degree radial. Illustrated in Figure 6-8, the navigation function will automatically turn to a 45-degree intercept heading when it finds the airplane is not already on the radial that you have dialed. As the airplane approaches the radial, the autopilot will slowly shallow the intercept angle. Upon reaching the radial, the autopilot will turn the airplane and track the radial.

### Flying an Assigned Heading to Intercept a Radial

Let's now try an even more complicated but very common intercept maneuver. Suppose you are en route to ECA and are instructed to fly a heading of 060 degrees, vector to intercept the 089-degree radial from ECA. You cannot simply dial the 089-degree radial and engage the navigation function as you did in the last example. Why? Because the autopilot will set you up on an intercept heading of about 45 degrees to intercept the radial. This doesn't work because ATC has specifically instructed you to fly a 30-degree intercept heading. One choice is to simply hand-fly the airplane on the 060-degree heading and then push the NAV button once you reach the 089-degree radial. Another choice is to take advantage of the following feature of the autopilot.

The navigation function can also be used in combination with the heading function to fly an assigned heading to intercept a radial. This procedure is illustrated in Figure 6-9.

Figure 6-9 shows how you can dial the heading you have been assigned to fly, twist the radial you wish to intercept, and then simultaneously press the HDG and NAV buttons. Looking at the flight mode annunciator as the airplane makes its way to the radial, it will seem that both the heading function and the navigation function are engaged at the same time. In fact, the heading function

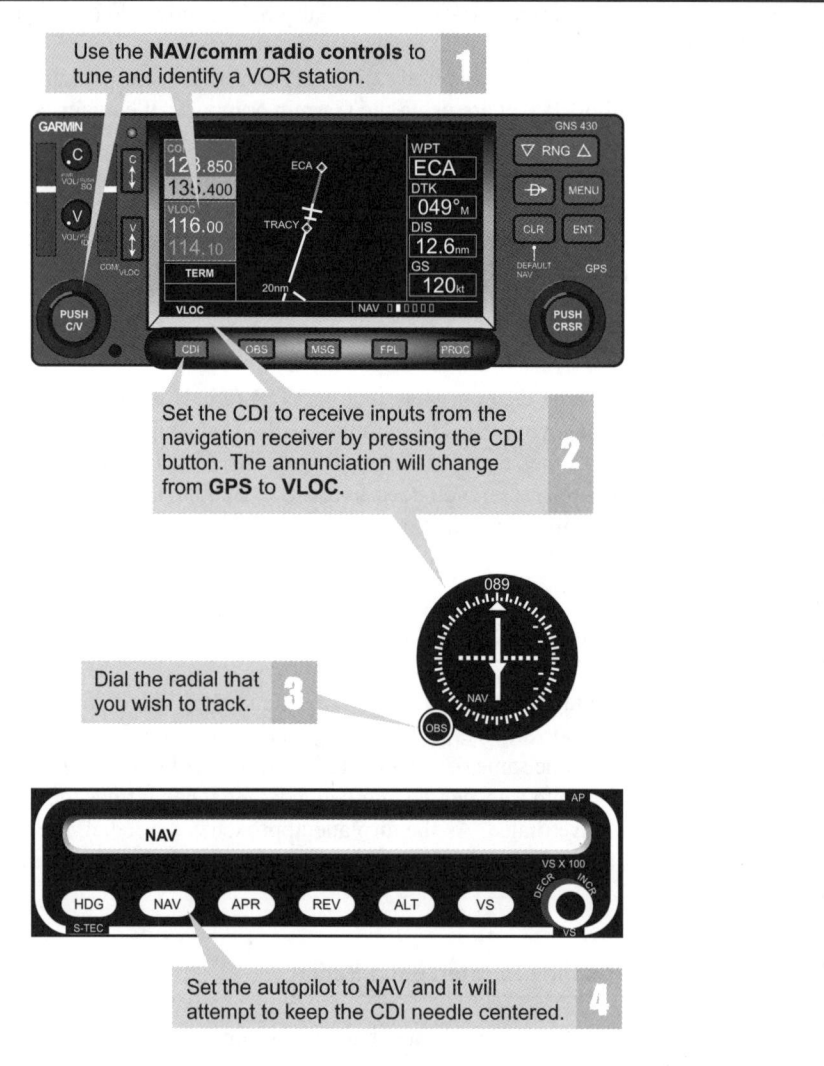

6.7. Tracking a VOR radial using the navigation function.

# 6 / Flying with an Autopilot

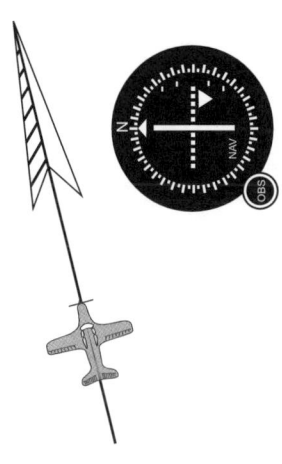

5. The autopilot will turn the airplane to a 45-degree intercept heading for the dialed radial.

6. Once the radial has been captured, the autopilot will turn the airplane to track the radial and attempt to keep the CDI needle centered.

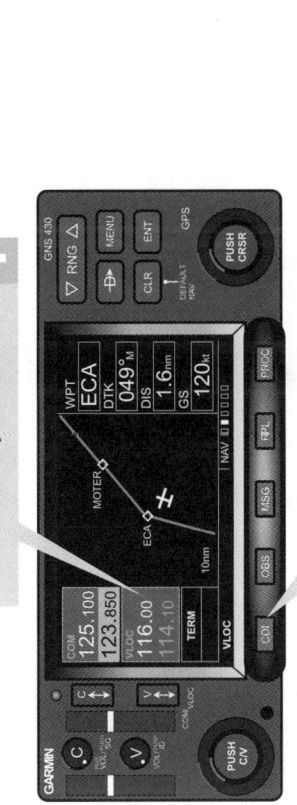

1. Tune and identify a VOR station.

2. Set the CDI to receive inputs from the navigation receiver by pressing the CDI button.

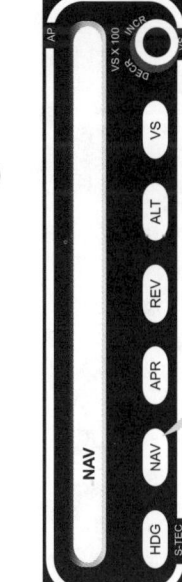

3. Dial the radial that you wish to track.

4. Press the NAV button on the autopilot control panel.

**6.8. Intercepting and tracking a VOR radial.**

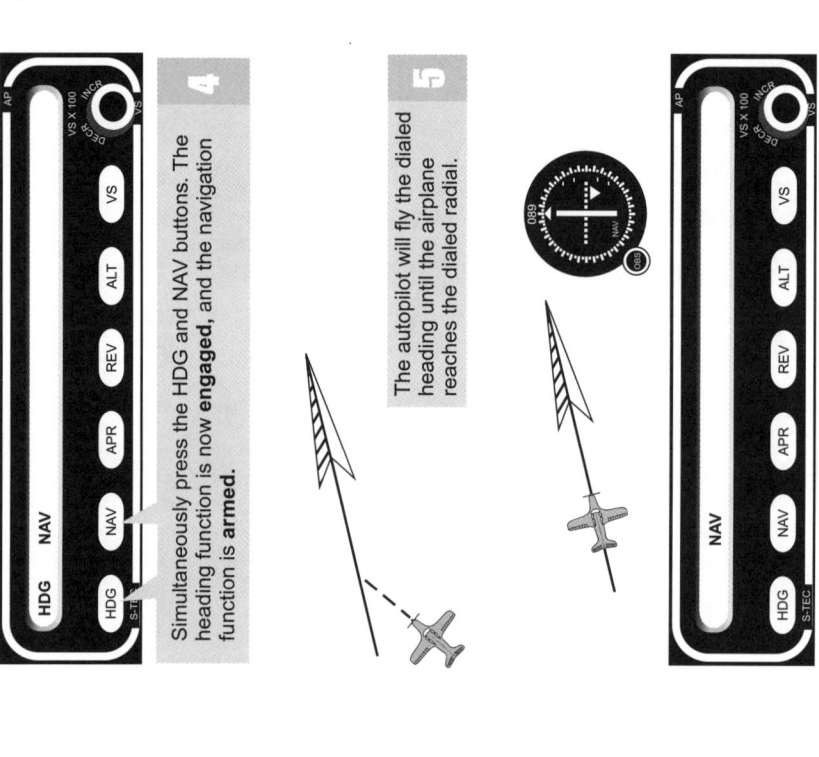

6.9. Flying an assigned heading to intercept a VOR radial.

is **engaged** while the navigation function is **armed.** An armed function is one that is set to engage once some condition is met. In this case, the navigation function is set to start performing the job of tracking the dialed radial once the heading function gets the airplane over to the radial. Once the airplane reaches the radial, the autopilot will automatically disengage the heading function, and the navigation function will then automatically switch from armed to engaged.

### Flying a Nonprecision Approach

Suppose upon intercepting the 089-degree radial, ATC clears you for the VOR approach into Oakdale. One choice is to use the autopilot's navigation function to continue tracking the radial outbound on the approach. Press the NAV button on the autopilot and engage the navigation function. You will notice that the navigation function provides you with the same kind of tracking accuracy that you see while tracking a VOR radial en route. If you remain within 4 NM of the airway, you are considered to be on course. However, flying an approach typically demands a much higher degree of accuracy. For this reason, autopilots typically offer a similar but more accurate function that tracks a navigation signal more closely.

The **approach function** is similar to the navigation function, with one important difference. The approach function has higher standards of accuracy. Use the approach function during the approach phase when accuracy is paramount. To engage the approach function, simply press the APR button on the autopilot control panel after you have already engaged the navigation function. You will now see APR appearing on the flight mode annunciator in addition to NAV, as shown in Figure 6-10. In a sense, you have engaged the navigation function and then added the approach function precision to it.

When executing a localizer nonprecision approach, you will find there is no need to press the APR button at all. The autopilot will sense that you have dialed a localizer frequency, and it will automatically engage the approach function after you press the NAV button. Using the approach function to capture and track a localizer is illustrated in Figure 6-11.

As with the navigation function, the approach function can also be used together with the heading function to fly an assigned heading to intercept a final approach course. As with using the heading and navigation functions, using the heading and approach functions together is a simple matter of simultaneously pressing the HDG and APR buttons. The procedure is illustrated in Figure 6-12.

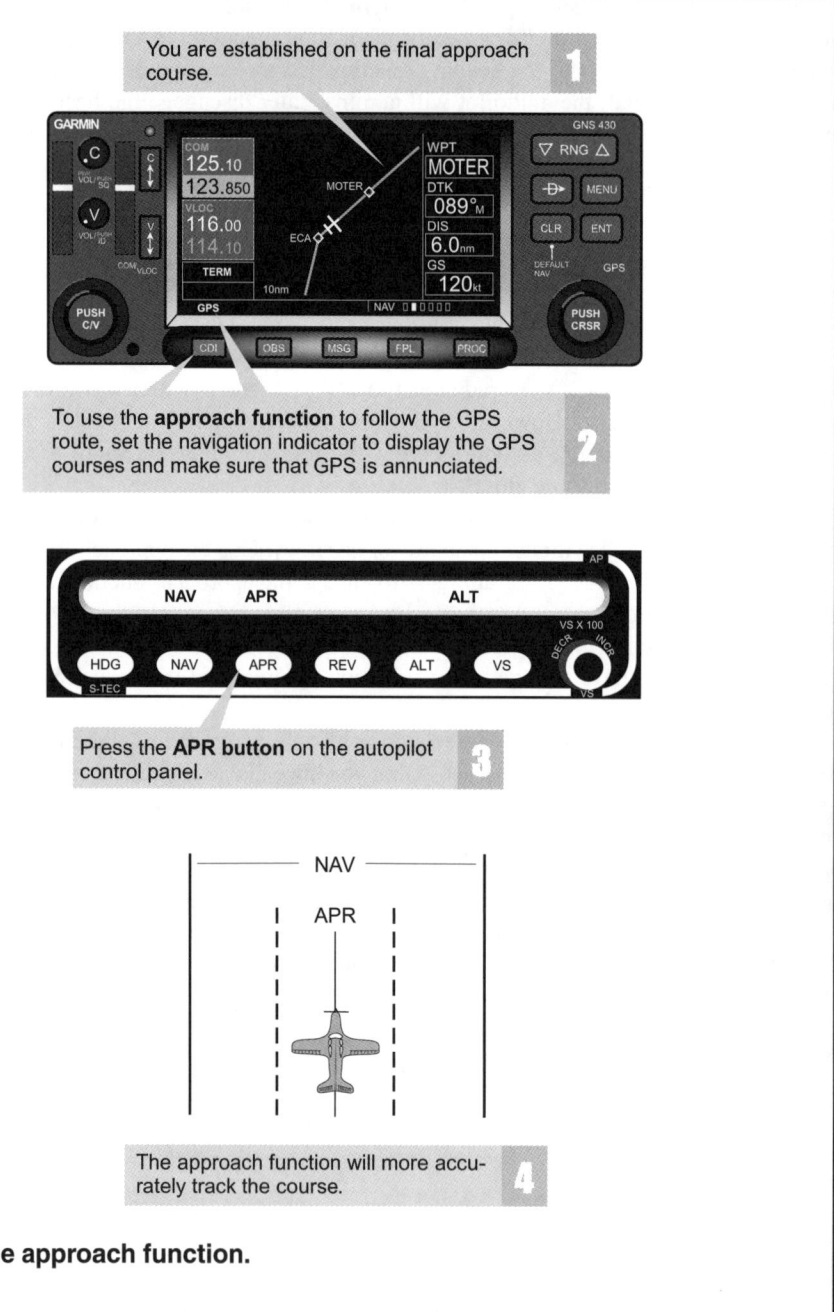

1. You are established on the final approach course.

2. To use the **approach function** to follow the GPS route, set the navigation indicator to display the GPS courses and make sure that GPS is annunciated.

3. Press the **APR button** on the autopilot control panel.

4. The approach function will more accurately track the course.

6.10. The approach function.

# 6 / Flying with an Autopilot

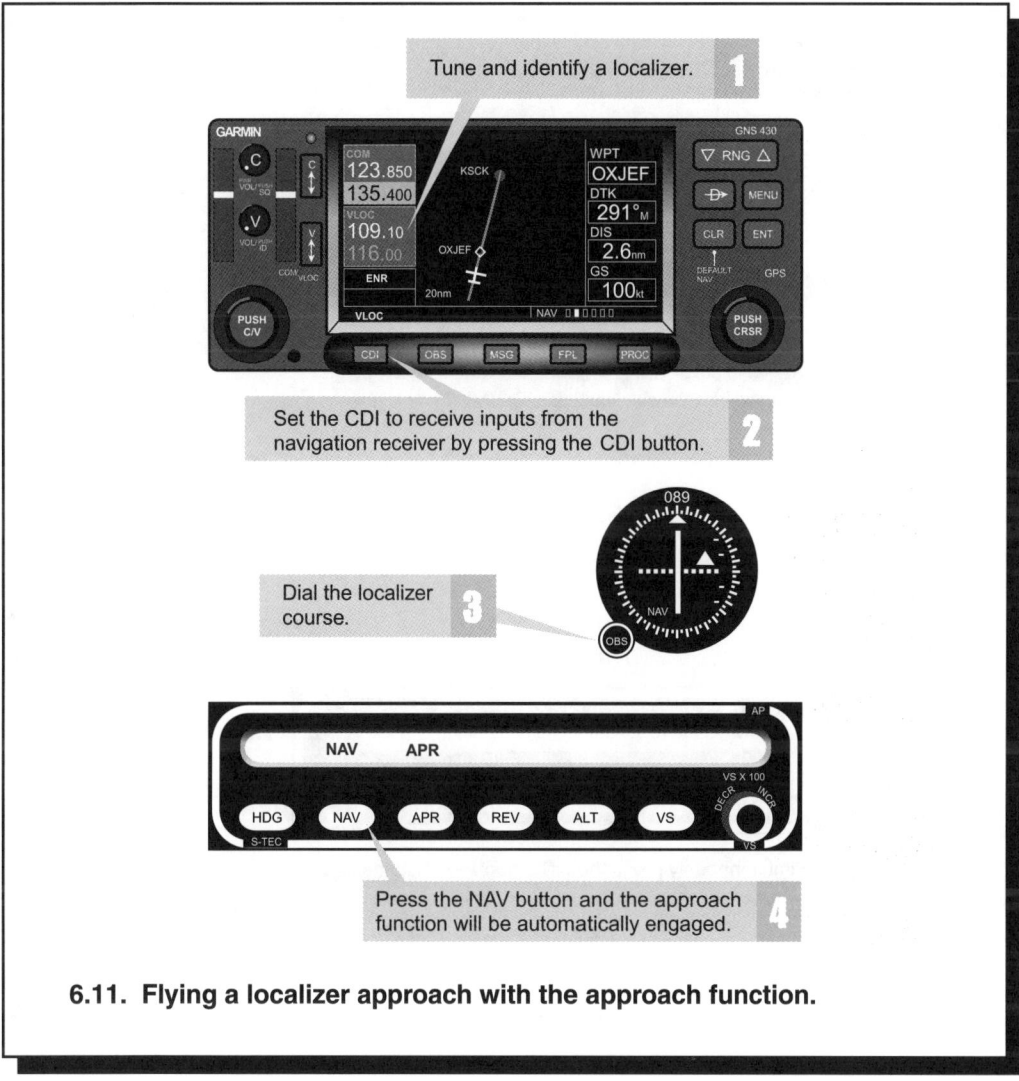

6.11. Flying a localizer approach with the approach function.

Set the computer to give you guidance to the final approach course (OBS mode is used here). **1**

Dial the assigned heading using the heading bug. **2**

Simultaneously press the HDG and APR buttons. **3**

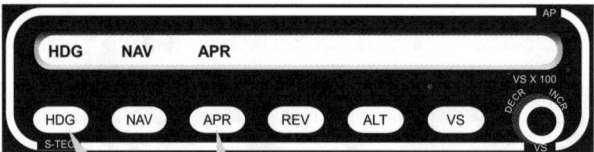
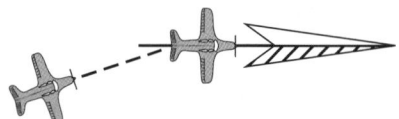

The autopilot will follow the dialed intercept heading until reaching the approach course. The approach function will then switch from armed to engaged, and the autopilot will turn to track the approach course. **4**

6.12. Flying an assigned heading to intercept an approach course.

## Flying a Precision Approach

As a final scenario, suppose you work your way down to the MDA at Oakdale and don't see the runway environment. You have planned nearby Stockton as your alternate airport where a precision ILS approach awaits you. You advise ATC of your intentions and get vectored to the final approach course at Stockton.

For the ILS approach into Stockton, in addition to tracking a localizer, you also have to track a glide slope. Once again, the autopilot can help you out here. Figure 6-13 illustrates how the approach function and the altitude can be used to fly an ILS approach.

In addition to engaging the approach function as you would for any instrument approach, you must also press the ALT button. Be sure to watch the flight mode annunciator as you progress toward the localizer and then the glide slope. You will see a succession of mode annunciations indicating that the autopilot is armed to capture both the localizer and the glide slope.

Once the localizer and glide slope have been captured, you now have the important task of managing the speed of the airplane using the throttle lever(s). Keep in mind that the autopilot is now using pitch to maintain the airplane on the glide slope. Your airspeed now is solely a function of the power setting that you choose. Spend some time learning what power settings result in what airspeeds and choose some sensible ones. Be sure to always monitor your position on the ILS approach as well as your airspeed at all times during this kind of **coupled approach.**

## No Vertical Autopilot Control Without Lateral Autopilot Control

It is important to keep in mind one rule about using the autopilot functions we have just discussed. In order to use any of the vertical autopilot functions such as altitude or vertical speed, you must first engage one of the lateral autopilot functions such as navigation, heading, or approach. Attempting to press the ALT or VS buttons on the autopilot control panel without first engaging one of the lateral autopilot functions will have no effect: Neither function will engage.

**1** Tune and identify the localizer frequency.

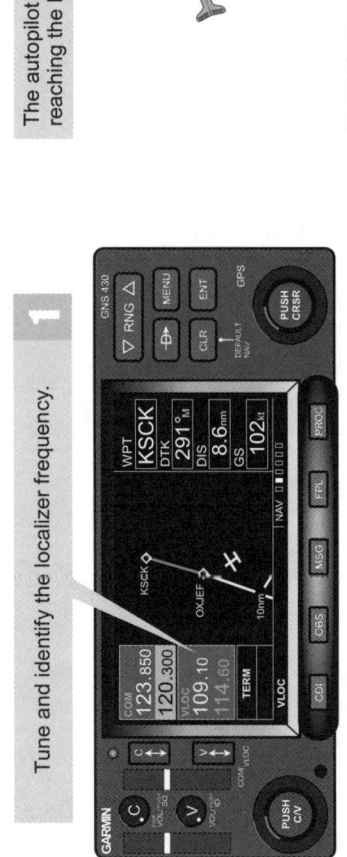

**2** Dial the assigned heading using the heading bug. Twist the localizer course using the OBS knob.

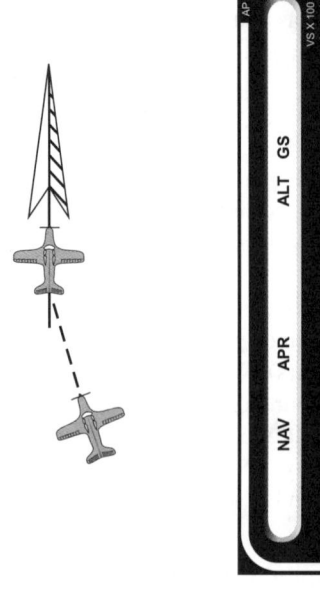

**3** Simultaneously press the HDG and NAV buttons. The approach function will arm automatically.

**4** The autopilot will follow the dialed intercept heading until reaching the localizer, then turn to track the localizer.

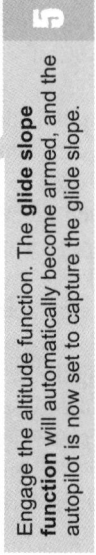

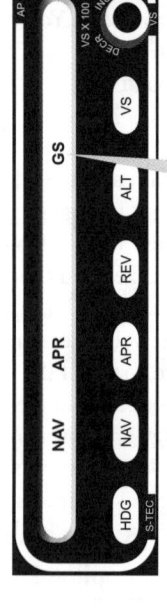

**5** Engage the altitude function. The **glide slope function** will automatically become armed, and the autopilot is now set to capture the glide slope.

**6** When the glide slope is captured, the glide slope function will become engaged.

**6.13. Flying a precision approach with the approach and glide slope functions.**

120

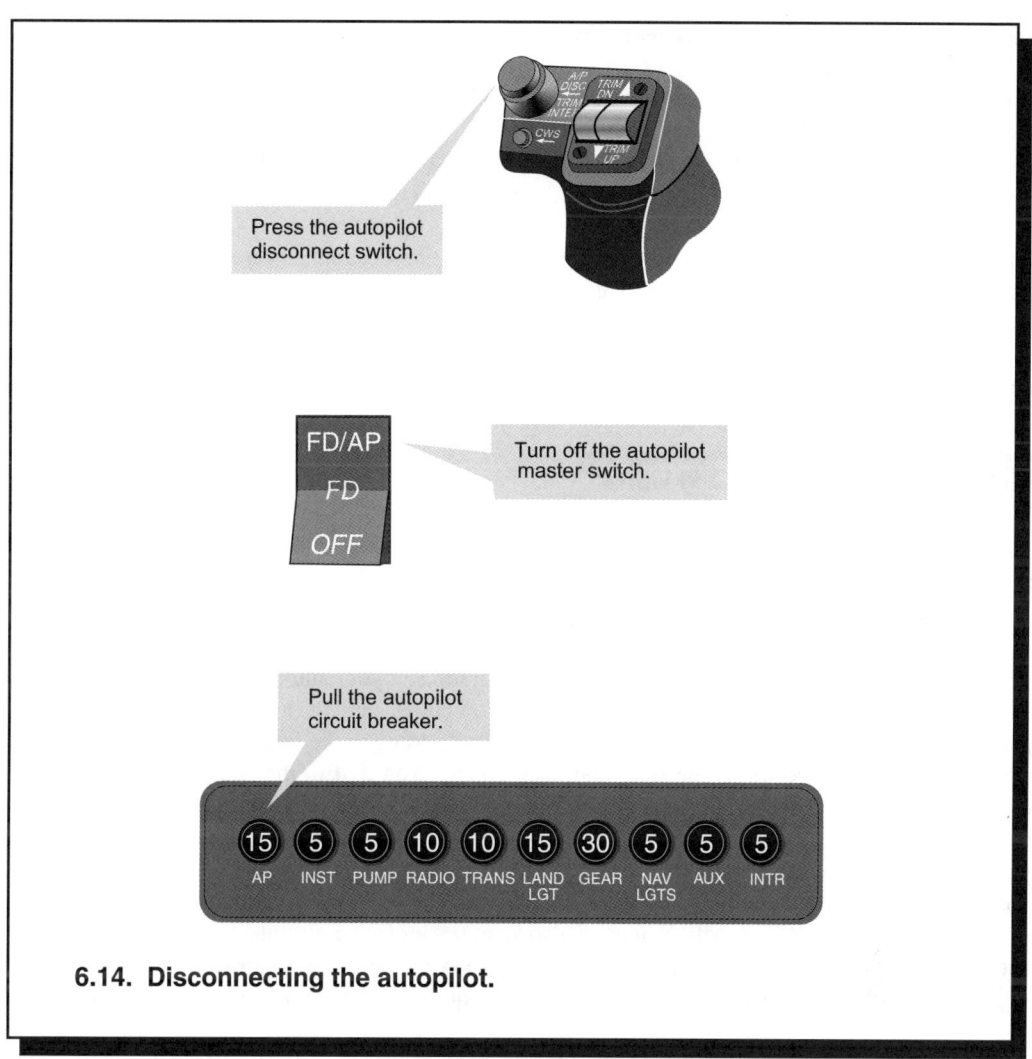

**6.14. Disconnecting the autopilot.**

### Disconnecting the Autopilot

You should become very familiar with the procedures required to disconnect or disable your autopilot. Most autopilots have several autopilot disconnect switches. One popular location for an autopilot disconnect switch is the control yoke (see Figure 6-14). You can disconnect the autopilot by switching off the master autopilot switch. Your airplane also has a circuit breaker that interrupts power to the autopilot. Lastly, autopilots installed in airplanes of any size can be manually overpowered. If attempts to disengage the autopilot and the servomechanisms fail, simply force the control yoke in the desired direction.

### Positive Exchange of Controls

As you remember from your early pilot training, your instructors likely emphasized the importance of knowing who is responsible for manipulating the controls of the aircraft at all times. When control of the aircraft is transferred between student and instructor, it was important to verbally acknowledge this act using phrases such as: "My controls," or "I've got the airplane." Following these procedures, it is less likely that there will ever be any doubt about who is flying the airplane.

Using an autopilot presents this same opportunity for confusion. Avoid the confusion by thinking of the autopilot as another pilot in the cockpit. When you opt to engage the autopilot, simply announce that the autopilot is being engaged and what autopilot function is being used. Here is a typical scenario to illustrate the concept. Suppose you are proceeding along an airway and engage the navigation and altitude functions. A useful announcement might be as follows: "Autopilot is coming on, navigation and altitude." You might also remind yourself that the autopilot will follow the desired track to the active waypoint, while you must check to see how you are doing along the airway given the winds.

## MORE ABOUT MODE CONFUSION

When using an autopilot in addition to a GPS computer, you now have several cockpit automation devices that can each operate in different modes. The possibilities for the kind of mode confusion we discussed in previous chapters are now even greater.

Suppose you are en route to an initial approach fix where you will commence a GPS approach. You have your GPS computer set in the sequencing mode and your autopilot's navigation function engaged. The GPS computer is providing the navigation targets, and the autopilot is steering the airplane to the targets. During your descent, instead of proceeding to the initial approach fix, let's assume you now get a vector to the final approach course. You dial in the heading you have been instructed to fly and set your autopilot to heading function. You now also set your GPS computer to the nonsequencing mode, make the FAF the active waypoint, and dial in the approach course. Once you intercept the approach course you have two mode changes to remember. You must reengage your GPS computer to the sequencing mode and then either reengage the autopilot to the navigation or approach function or turn it off. You must also keep track of your altitude and make sure that the autopilot levels off at 3,000 feet prior to intercepting the final approach course.

Figure 6-15 illustrates many things you will be required to keep track of as you complete this approach.

Looking at Figure 6-15, you can see that staying on top of what modes are in charge is no easy task. This is perhaps the most challenging aspect of learning to be proficient in the modern automated cockpit. You will learn that you must include the many mode annunciations in your scan and remain aware of what your computers are doing now and planning to do next. It is often said that cockpit automation systems were designed to make your job easier. You won't be the first pilot to wonder if cockpit automation doesn't sometimes make your job harder.

## PREFLIGHTING AND MONITORING YOUR AUTOPILOT

More than any other automated system in the modern general aviation cockpit, the autopilot is the one with the most interconnected parts and the most vulnerable to unexpected behaviors. It is particularly important to preflight check your autopilot prior to departure and to monitor the operation of your autopilot at all times during your flight. After flying with several different kinds of autopilots, I have watched them make uncommanded turns, climbs, and dives; drift from a dialed heading, course, or altitude; and disconnect, all for unexplained reasons.

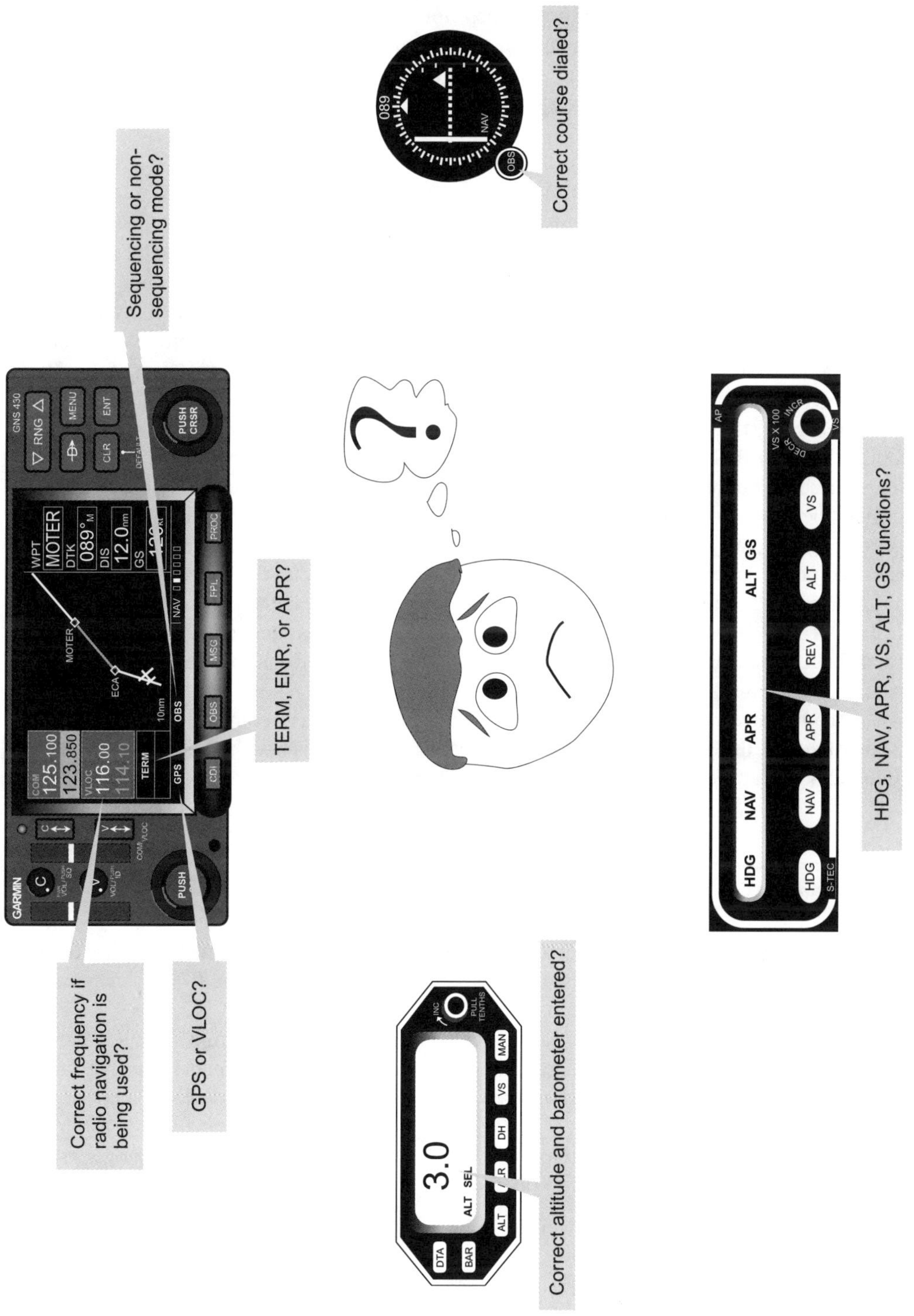

# 6 / Flying with an Autopilot

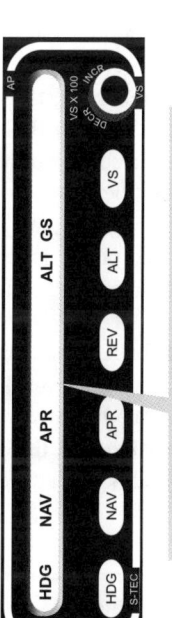

**6.15. Mode confusion.**

## Preflight Checks

The pilot operating handbook (POH) supplement for your autopilot will contain a procedure for a preflight check that must be performed prior to departure. This preflight check will help you discover some of the problems that your autopilot might be experiencing. Learn the preflight procedure for your autopilot and do it prior to every flight.

## In-Flight Monitoring

Even if your autopilot passes the preflight check, it is important to stay on top of what it is doing at all times to guard yourself against unexpected behaviors in flight. One approach I find useful is to think of my autopilot as a student pilot. The student is capable of carrying out many useful tasks but can always be counted on to do something unexpected every once in a while. I use my autopilot regularly but keep a watchful eye on it at all times. Don't get caught napping when your cockpit automation decides to do something crazy. The human pilot is an important part of the design of every cockpit automation system. When the system does something unexpected, the role of the pilot is to recognize the situation, disconnect the automation, and make the situation right again.

Another kind of in-flight situation that requires pilot intervention happens when the autopilot is functioning correctly, but the circumstances change such that using the autopilot is no longer desirable. Consider the case of a gradual, unnoticed partial power loss during cruise flight. With the altitude function engaged, the job of the autopilot is to manipulate pitch to maintain the dialed cruising altitude. As the power is reduced, the autopilot will continue to pitch up to maintain the altitude. Meanwhile, airspeed gets slower and slower. If this situation continues, the airplane will continue to slow down all the way to the stall speed. In this situation, the autopilot is functioning properly. What is abnormal about the situation is the pilot's decision to continue using the autopilot. Icing is a more common example that results in this same situation.

Again, you must always think of the pilot and cockpit automation as a combined system. In our present state of the art, the cockpit automation systems discussed in this book are useless without a human pilot taking an active part in the process of flying the airplane.

## SOME OTHER AUTOPILOTS

Let's take a quick look at two other popular autopilots. The autopilot illustrated in Figure 6-16 offers capabilities similar to the one we discussed earlier. You can see that aside from a few minor design differences, this autopilot functions in the same way.

The autopilot illustrated in Figure 6-17 is a more sophisticated system offering several features that less expensive systems do not offer. The more informative flight mode annunciator, indicated air speed (IAS), and go-around (GA) functions are features found on the flight decks of modern jet transport airplanes.

More details about these and other autopilot systems can be found in the manufacturer's pilot information manuals available on the World Wide Web at

www.bendixking.com

# 6 / Flying with an Autopilot

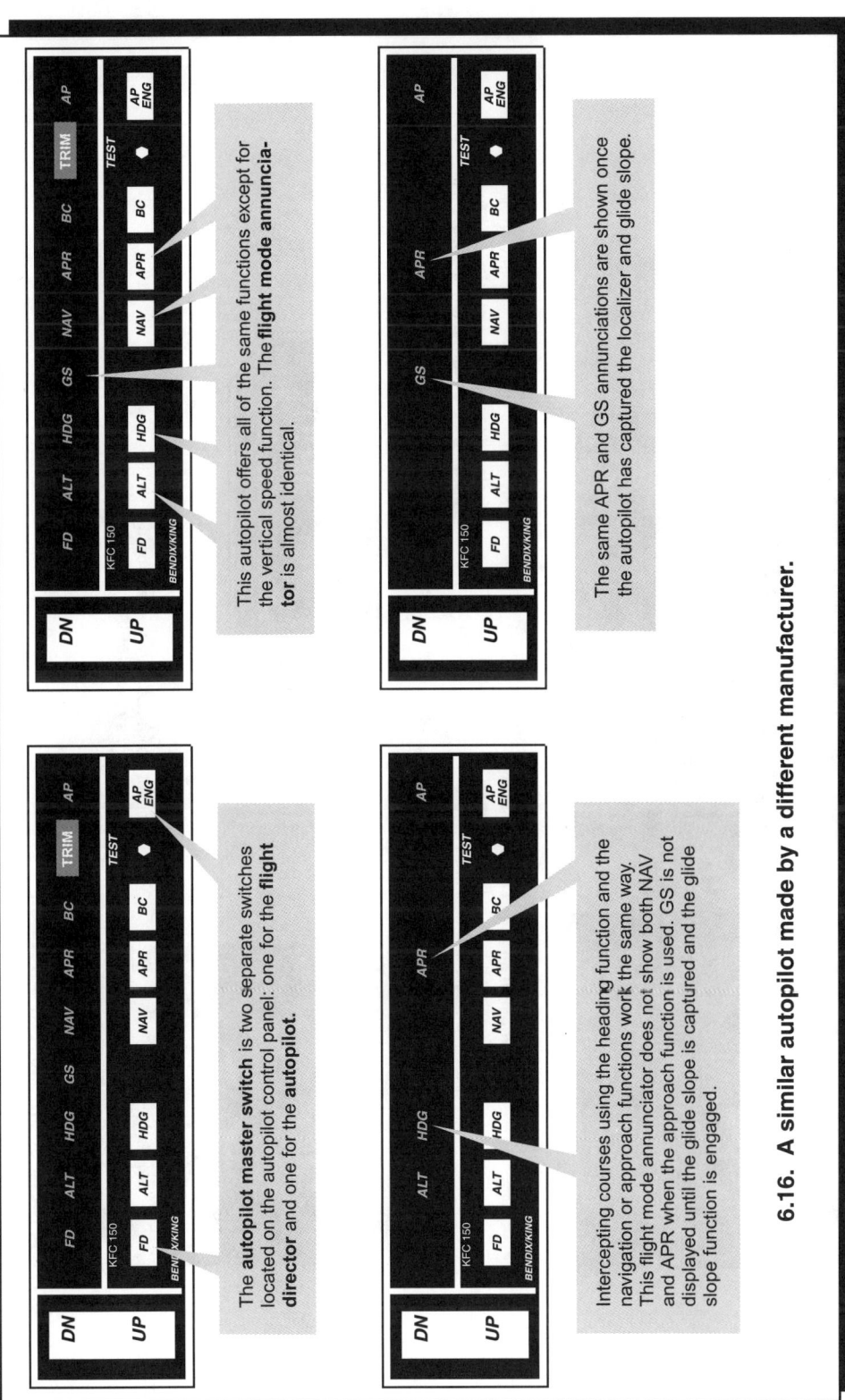

6.16. A similar autopilot made by a different manufacturer.

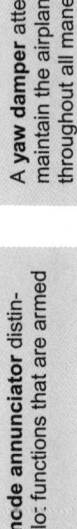
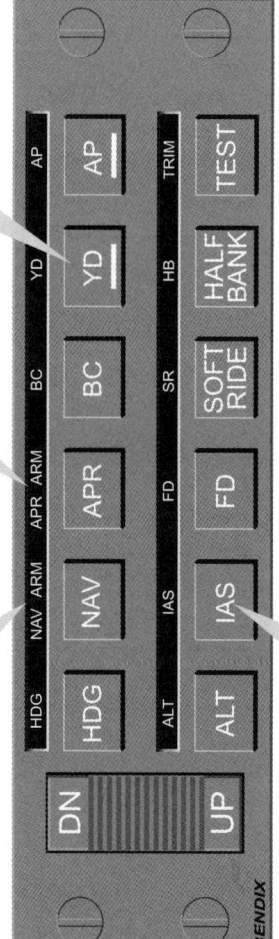

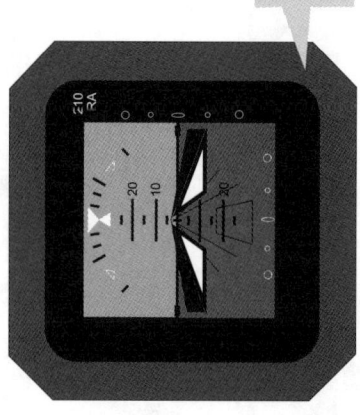

A more detailed **flight mode annunciator** distinguishes between autopilot functions that are armed and engaged.

A **yaw damper** attempts to automatically maintain the airplane in coordinated flight throughout all maneuvers.

A separate **altitude selector** must be installed to use the **vertical speed function**. No VS annunciation appears on the autopilot flight mode annunciator.

The **indicated airspeed (IAS) function** uses pitch to perform constant-speed climbs, cruises, and descents.

This autopilot interfaces with an **electronic flight instrument system (EFIS)**.

6.17. A more sophisticated autopilot.

 **PRACTICE SESSION (AIRPLANE)**

If you've worked hard and made it this far, here's your payback. This practice session was consistently reported to be the most fun by all of the pilots with whom I flew. Engaging the navigation and altitude functions for the first time and watching the airplane fly itself was quite an experience for most. For the pilots who aspired to fly for an airline, I couldn't help asking, "Are you sure you want to do this for a living?"

Be sure to read the supplement in your airplane's POH for the autopilot you are using and familiarize yourself with the preflight (and in-flight) procedures that the manufacturer prescribes for its product. Read the manufacturer's pilot information manual for your autopilot.

Plan to fly to an airport where you can perform several different kinds of instrument approaches. Ideally, find one that offers a GPS approach, a VOR or localizer approach, and an ILS approach.

**Skills to Practice**

1. Use the autopilot's navigation function once you reach a safe maneuvering altitude and turn to track your programmed flight route.
2. Use the vertical speed function to climb to your assigned cruising altitude.
3. Use the altitude function once you reach your assigned cruising altitude.
4. Use the vertical speed function to execute a descent to a crossing restriction that you program with the GPS computer.
5. Use the approach function and vertical speed function to fly a simple GPS approach in the way you flew your first GPS approach.
6. Try a vectored GPS approach for which you have to engage the heading function to turn to an intercept heading and arm the navigation function to take over once you are established.
7. Try an ILS approach and use the approach and glide slope functions to track the localizer and glide slope signals.
8. Practice disconnecting your autopilot using all of the available methods in your airplane.

Use good judgment when deciding under what circumstances to use your autopilot. Just because your autopilot can track a localizer and glide slope all the way down to the runway surface doesn't mean you should use it this way. Choose sensible minimums for yourself. Also be sure to check your instrument approach plates for any restrictions on the use of autopilots during approaches. Coupled approaches are not approved at some airports.

 **DVD DEMONSTRATION DISC**

The use of the autopilot during a complete flight is demonstrated on the accompanying DVD disc (Skill 18).

# CHAPTER 7

# Avoiding Traffic, Terrain, and Weather

*In this chapter, you will become familiar with three other kinds of computers in the cockpit. These systems support you in avoiding terrain, traffic, and weather hazards en route. You will learn how to work with these systems in much the same way as you have learned to work with your GPS computer and autopilot. The new computers you will learn about offer you many interesting services. In exchange for these services, you will have to gain a basic understanding of how the systems work along with their capabilities and limitations. You will learn the purpose of these systems, which is to empower the vigilant pilot with additional information. Working together, the human pilot and automated systems can make a great team. The pilot who ceases to monitor the situation and leaves the automation to do it alone is asking for trouble.*

## TRAFFIC COLLISION AVOIDANCE SYSTEMS

A **traffic collision avoidance system** (**TCAS**) alerts you to threats posed by nearby aircraft. Mandatory equipment onboard all airliners since the mid-1970s, TCAS has recently become available to general aviation airplanes. Often referred to as the "fish finder" or the "metal detector," TCAS attempts to locate nearby airplanes and provide alerts and advisories. TCAS is a self-contained onboard system: It must be installed onboard an airplane and does not rely on any other equipment outside of the airplane.

How does TCAS locate other airplanes in the sky? TCAS works by querying the transponders of nearby aircraft to determine their distance, bearing, and movement relative to your aircraft. In addition, TCAS uses the Mode C information transmitted by transponders to determine the altitude and vertical movement of surrounding aircraft. Using these capabilities, TCAS can serve as somewhat of a miniature ATC radar facility, providing you with traffic alerts and advisories.

Not as capable as a human air traffic controller, the most sophisticated TCAS systems only provide limited services. All TCAS systems can issue a **traffic advisory (TA)** whenever another transponder-equipped aircraft appears to come within an approximate 40-second range of your aircraft. Traffic advisories take the form of an aural alert: Traffic! Traffic!

More advanced TCAS systems (TCAS II) can also issue a **resolution advisory (RA)** when another transponder-equipped aircraft comes within an approximate 25-second range of your aircraft. Resolution advisories take the form of an avoidance command that instructs you how to fly the plane in order to avoid the threat. You will hear an aural alert that instructs you to perform a vertical avoidance maneuver. Example aural alerts are "Climb! Climb!" or "Descend! Descend!"

Every TCAS has a **traffic display.** Figure 7-1 shows two kinds of traffic displays available for the small-airplane cockpit. The display at the top of Figure 7-1 is called a **dedicated traffic display.** This type of display is capable of showing only traffic information. The display at the bottom of Figure 7-1 is called a **multifunction display (MFD).** A multifunction display is capable of presenting many different kinds of information, including traffic, weather, terrain, route information, and more. Multifunction displays are becoming a popular choice as new kinds of cockpit automation systems are developed. A multifunction display eliminates the need to install a separate dedicated display for each new cockpit automation system installed in an airplane.

To see traffic information on the MFD shown in Figure 7-1, simply press the TRFC button on the front of the unit.

TCAS traffic displays all use the same symbology to present traffic information. Figure 7-2 shows several important kinds of symbols used on traffic displays. Note that the resolution advisory symbols only appear when an advanced TCAS II system is used.

Every TCAS system has a **TCAS control panel.** Figure 7-3 shows a simple kind of TCAS control panel. This TCAS control panel contains the typical controls for setting the range of the traffic display. The higher the range that is set, the more traffic you will see. However, the traffic symbols will appear more densely packed and difficult to distinguish. The altitude limit control performs the same function for vertical range. Different settings allow different amounts of airspace to be included on the display.

Figure 7-4 shows an advanced TCAS II control panel that is integrated with a dual transponder unit. Aside from the transponders, the main difference about this TCAS control panel is that it allows you to turn traffic and resolution advisories on and off. Pilots are to be cautioned about turning off TCAS functions: If you don't ask for help, you won't get any help.

Some advanced TCAS II systems have an **integrated TCAS vertical speed indicator.** This display is designed to simplify the process of executing avoidance maneuvers during a TCAS resolution advisory. One integrated TCAS/VSI display is shown in Figure 7-5.

Using a series of green and red lights, the integrated display shows you acceptable vertical speeds to fly during the maneuver. Upon hearing a resolution advisory such as "Climb! Climb!" you simply have to get the VSI needle out of the red region and into the green region.

# 7 / Avoiding Traffic, Terrain, and Weather

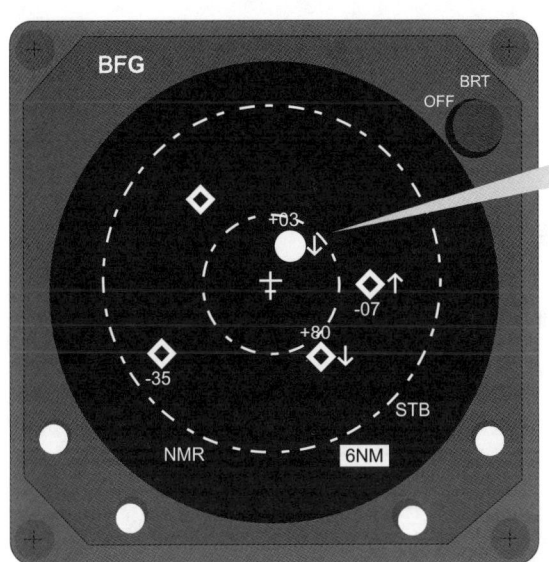

A **dedicated traffic display** presents symbols for traffic that indicate the position, relative altitude, vertical movement, and threat condition of surrounding aircraft.

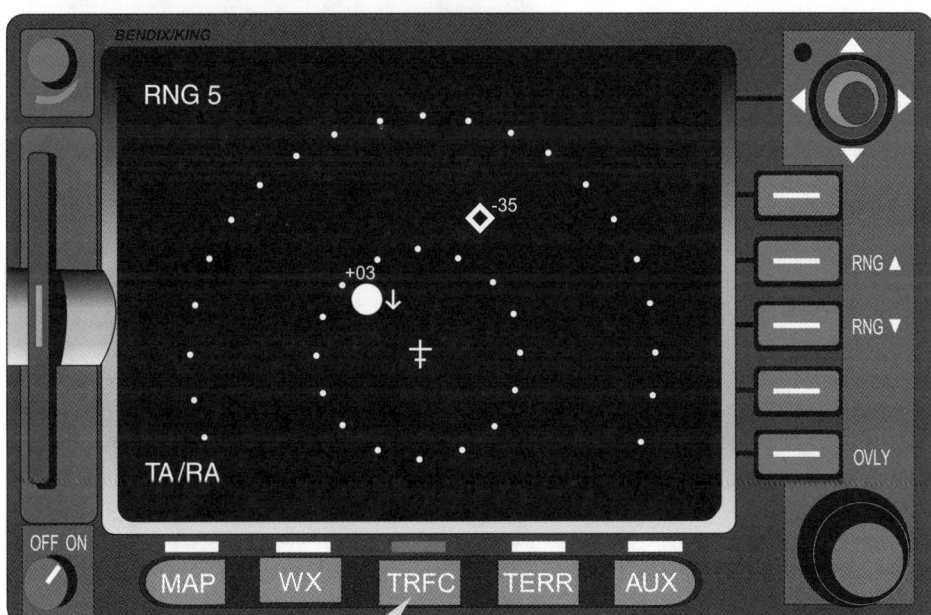

A **multifunction display (MFD)** is capable of presenting many kinds of information, including traffic. Press the TRFC button to see traffic information on this particular multifunction display.

**7.1. TCAS traffic displays.**

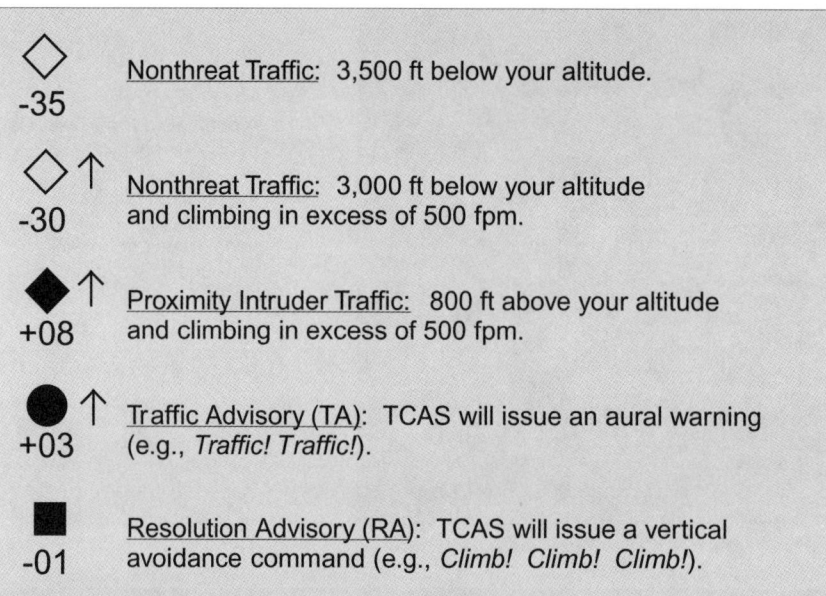

**7.2. Traffic display symbology.**

**7.3. TCAS control panel.**

# 7 / Avoiding Traffic, Terrain, and Weather

**7.4. TCAS II control panel.**

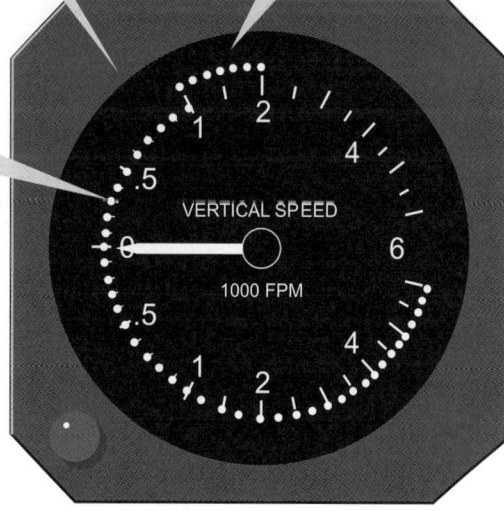

**7.5. Integrated TCAS and vertical speed indicator.**

An even more interesting integrated display is shown in Figure 7-6. This integrated TCAS vertical speed indicator additionally includes a traffic display. This display combines most of the TCAS information you will need during any traffic conflict situation.

Despite its many advantages, there are also several important limitations of TCAS. For example, TCAS cannot detect aircraft that do not have transponders. Another limitation of TCAS is that it often gives unwanted alerts when you are purposefully operating in the vicinity of other aircraft. For example, two aircraft making approaches to parallel runways will likely receive TCAS alerts. These alerts can serve as distractions while you are trying to perform this challenging maneuver. Perhaps the most undesirable aspect of TCAS is that some pilots tend to rely too much on TCAS. That is, there is sometimes a tendency toward complacency when automated systems are used. Users of TCAS might adopt the attitude: "Why do I need to look out for traffic now that I have this expensive computer to do it for me?" This reasoning is faulty. Because of its many limitations, TCAS is no substitute for a well-trained human pilot. Even when you have TCAS onboard your aircraft, you must consider it as a backup while you assume primary responsibility to see and avoid other aircraft.

This integrated TCAS/VSI display additionally includes a miniature traffic display.

**7.6. A more sophisticated integrated TCAS and vertical speed indicator.**

## GROUND PROXIMITY WARNING SYSTEMS

A **ground proximity warning system** (**GPWS**) alerts you to threats posed by surrounding terrain. The ground proximity warning system was designed to help reduce the number of **controlled flight into terrain** (**CFIT**) accidents. A ground proximity warning system works by comparing the position and movement of the airplane against a **terrain and obstacle database.** This database attempts to detail every obstruction that could pose a threat to an aircraft in flight. Whenever the GPWS determines that the airplane position, altitude, and trajectory place it near obstacles or terrain in the database, the system responds by issuing an alert.

The ground proximity warning system tracks the position and movement of the airplane using a global positioning system receiver. The airplane's transponder provides the ground proximity warning system with altitude information. Some systems additionally accept inputs from an outside air temperature probe to make altitude corrections for nonstandard temperature. These tracking systems, combined with the terrain database, help the GPWS to determine the situation of the aircraft at all times.

There are two basic ways in which a GPWS can support you in maintaining awareness of the aircraft's position with respect to surrounding terrain.

**Aural alerts** warn you about specific situations that present a terrain collision hazard. Using a "look ahead" function based on the aircraft's ground speed, the GPWS is able to alert you to upcoming terrain. At a closure time of approximately one minute, a "Caution! Terrain!" alert is issued. This alert changes to the more serious "Terrain! Terrain! Pull up!" alert when the closure time reaches 30 seconds. A second type of aural alert warns you about excessive descent rates sensed by the GPWS ("Sink Rate!" or inadvertent loss of altitude after takeoff ("Don't Sink!").

A **terrain display** allows the ground proximity warning system to present you with a pictorial view of surrounding terrain. This allows you to monitor the position of the aircraft with respect to surrounding terrain at all times. Figure 7-7 shows terrain displayed on a multifunction display.

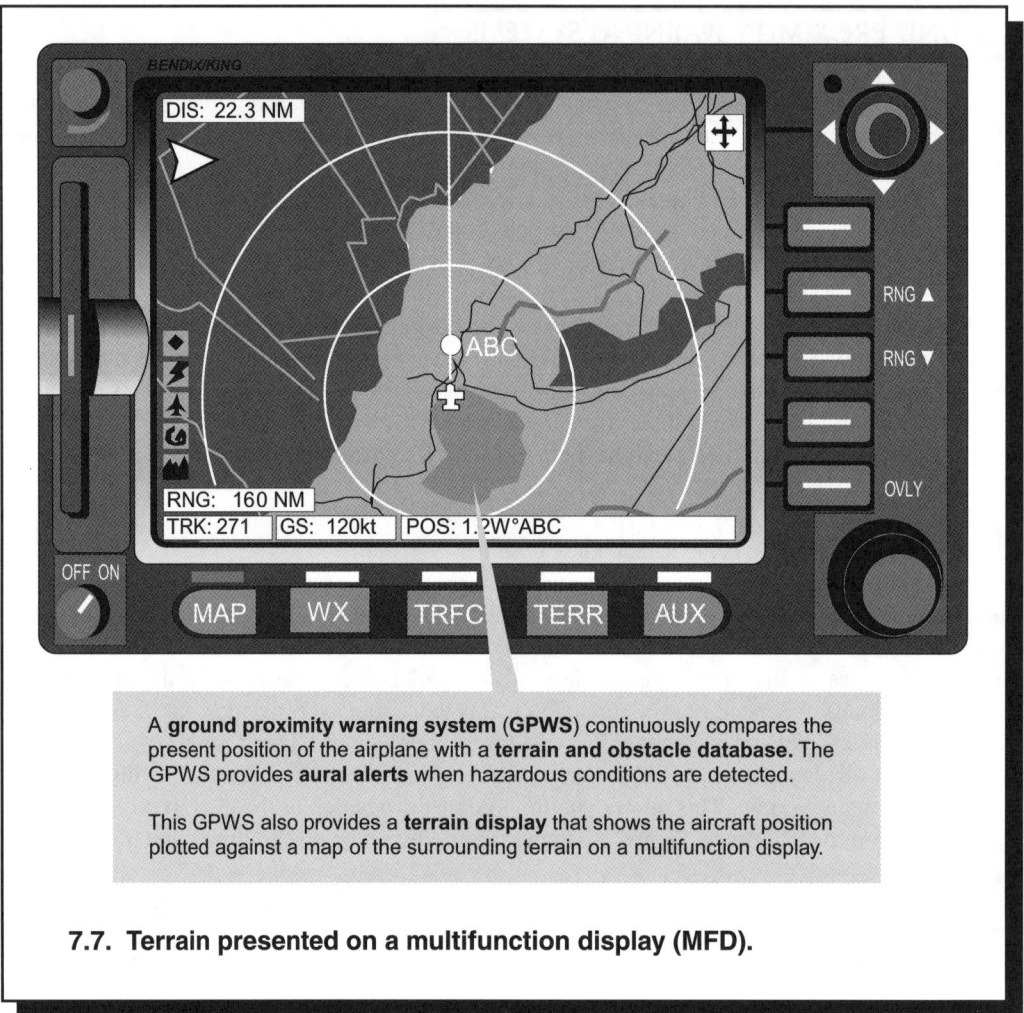

**7.7. Terrain presented on a multifunction display (MFD).**

Figure 7-8 shows the familiar color-coding conventions used on terrain displays to portray the relative threat to the aircraft based on the aircraft's present altitude.

The introduction of ground proximity systems has sharply reduced the number of CFIT accidents. Despite this significant leap forward in safety, incidents and accidents involving terrain still happen. In the modern GPWS-equipped cockpit, some of these incidents have been related to pilots' reactions to GPWS alerts. Not a perfect and all-knowing system, GPWS sometimes gives alerts when you deem them to be not necessary. These are what many pilots refer to as **nuisance alerts.** Most GPWS systems contain software logic that attempts to recognize and remain silent in situations in which proximity to terrain is normal. This logic is partly based on the airplane's distance

# 7 / Avoiding Traffic, Terrain, and Weather

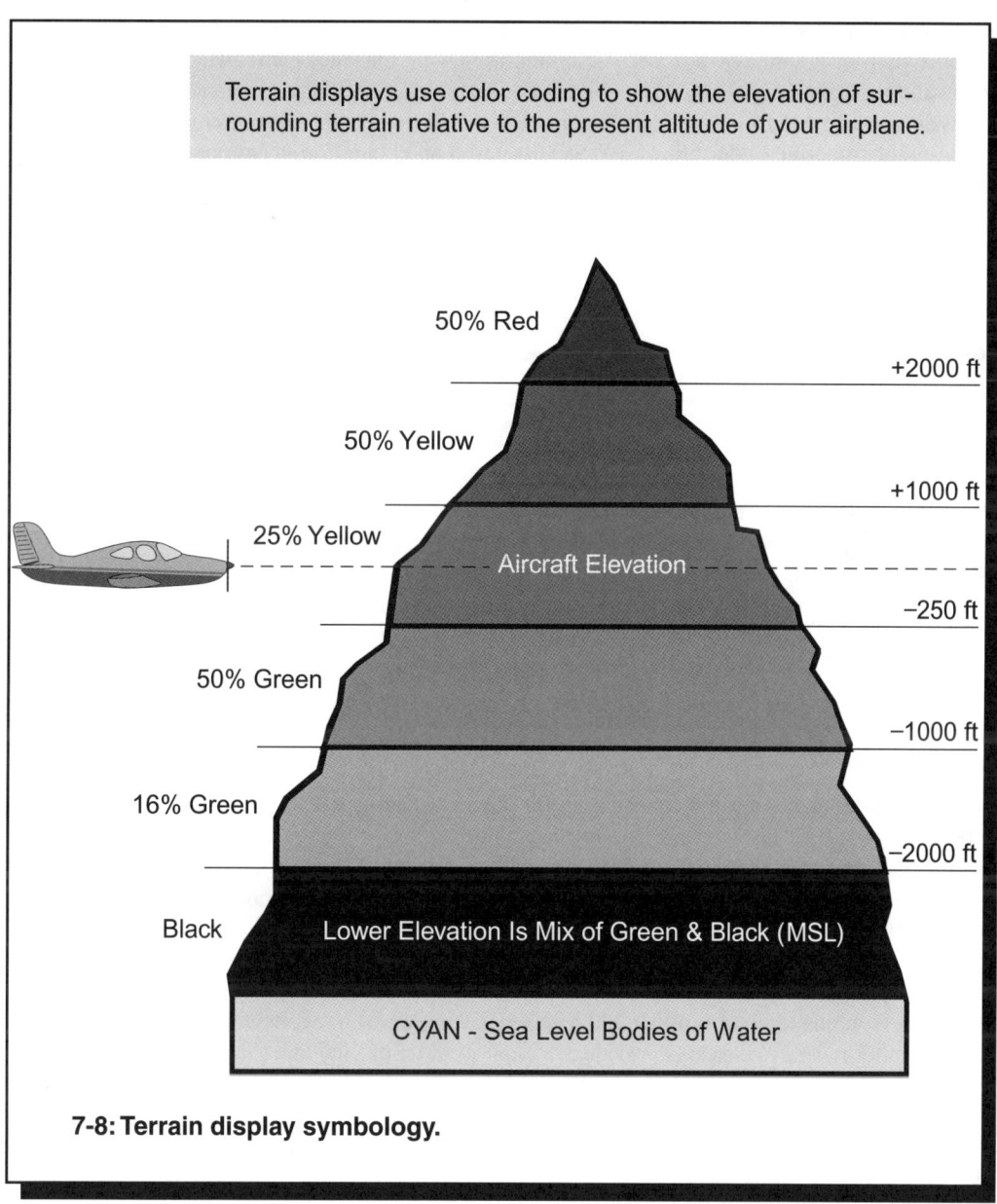

**7-8: Terrain display symbology.**

from the runway of intended landing. For example, flying at an altitude of 200 feet above ground level (AGL) when 3,500 feet away from the runway is reasonable. Flying at an altitude of 200 feet AGL when 5 miles from the runway is not reasonable. The GPWS logic attempts to remain silent in normal situations and speak up in abnormal situations. Despite the system's logic, nuisance alerts still sometimes happen. Consequently, most GPWS systems offer a **terrain inhibit switch** that allows you to silence alerts given by the GPWS. There have been cases in which pilots have used the inhibit switch or ignored GPWS alerts, thinking they were nuisance alerts, when in fact the alerts were valid indications of a dangerous situation. For this reason, it is recommended that pilots train themselves to respond to GPWS alerts just as they would to any other sort of emergency. The practice of simply ignoring or disabling GPWS alerts based on pilot intuition has not proven to be a safe one. Your manufacturer's reference manual and airplane flight manual supplement will prescribe specific procedures for responding to GPWS alerts. It is a good idea to learn and practice these procedures regularly.

## AIRBORNE WEATHER DETECTION SYSTEMS

To aid the pilot in avoiding hazardous in-flight weather, two types of onboard weather detection systems now exist for general aviation airplanes. Both systems are primarily designed to aid in avoiding thunderstorms and the hazardous flight conditions associated with thunderstorms such as hail and turbulence. These two systems operate using completely different principles, and each have their own unique capabilities and limitations. For this reason, many airplanes are equipped with both kinds of systems. It is important to keep in mind that both types of airborne weather detection system are only able to detect thunderstorms and their associated hazards. Neither type of system is generally able to detect clouds, fog, turbulence, icing, or IFR conditions.

### Weather Radar Systems

A **weather radar system** transmits radio pulses ahead of the airplane along its flight path. When these radio pulses encounter significant masses of water droplets, they are reflected back to the airplane. The strength of a reflection (also called an **echo** or **return**) indicates the size and density of the weather phenomenon. Large, dense masses of water droplets produce strong reflections, while smaller masses of droplets produce weaker reflections. The time it takes for the radio signal to be reflected back to the airplane determines the distance between the airplane and the water mass. These two principles give RADAR its name: RAdio Detecting And Ranging.

Aside from the radio transmitter and receiver that are a part of every weather radar system, airborne weather radar systems also provide a cockpit weather display that shows radar reflections returned from the area ahead of the airplane.

Airborne weather radar concepts are illustrated in Figure 7-9.

Becoming a proficient operator of a weather radar system is somewhat of a learning task. Some manufacturers of weather radar systems offer training courses to help customers make better use of their products. Other weather radar experts offer similar training seminars. The book *Airborne Weather Radar: A User's Guide* by James Barr (Iowa State University Press) gives an excellent introduction to the use of weather radar systems.

**Airborne weather radar systems** transmit radio pulses in front of the airplane and sense reflections from weather phenomena.

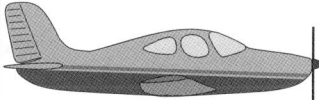

Airborne weather radar systems detect masses of water droplets, calculate the distance and bearing of these weather phenomena, and present them on a **cockpit weather display**.

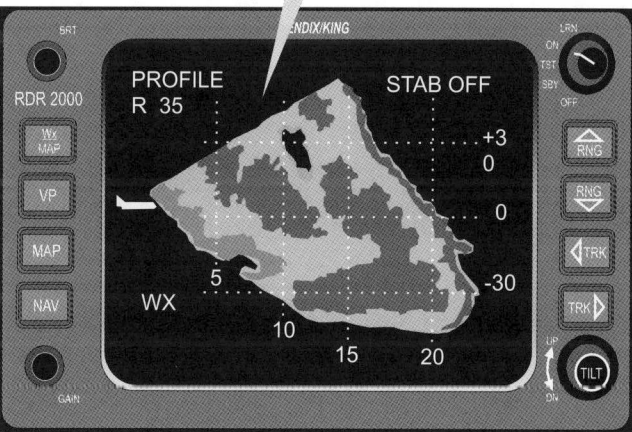

**7.9. Airborne weather radar system.**

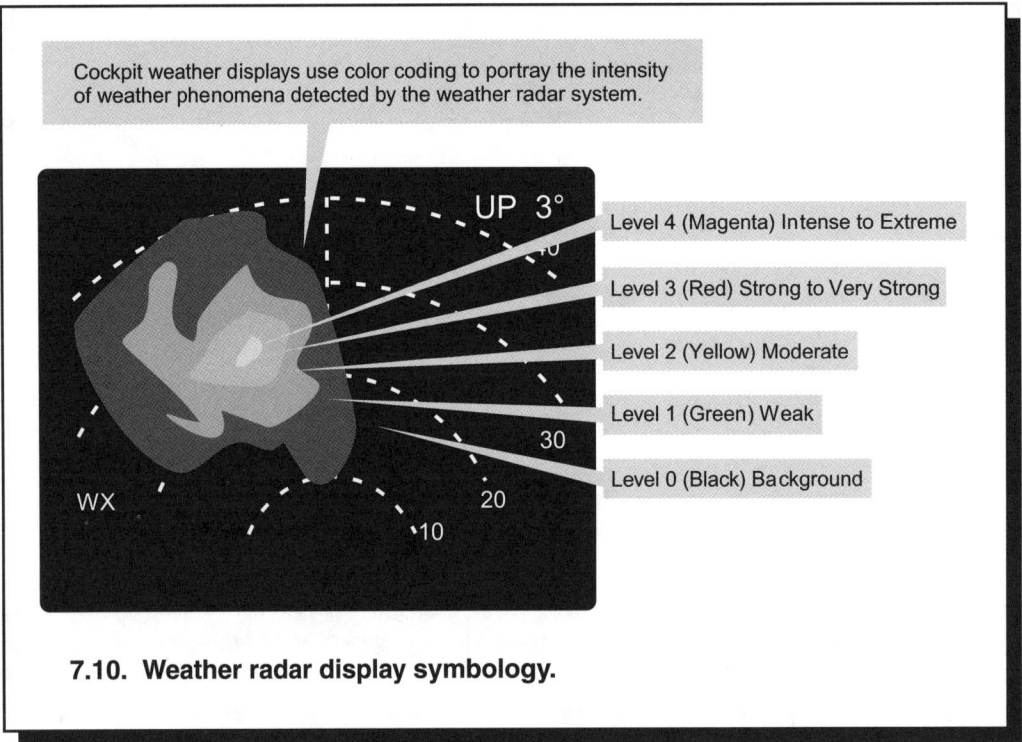

7.10. Weather radar display symbology.

*Interpreting the Radar Image*

A first step in learning to use weather radar is interpreting the images you see on the weather display. Figure 7-10 shows the color-coding scheme used by most weather radar systems to depict the intensity of radar echoes.

The changes in color across a cell appearing on the radar display provide clues about the nature of that cell. Large changes in intensity across short distances often indicate severe turbulence. The radar display in Figure 7-11 shows two cells. The cell to the left has a sharp gradient, suggesting the presence of severe turbulence.

The shapes of radar echoes can often give clues about the presence of hail. Figure 7-12 shows four differently shaped cells that suggest the possibility of hail.

# 7 / Avoiding Traffic, Terrain, and Weather

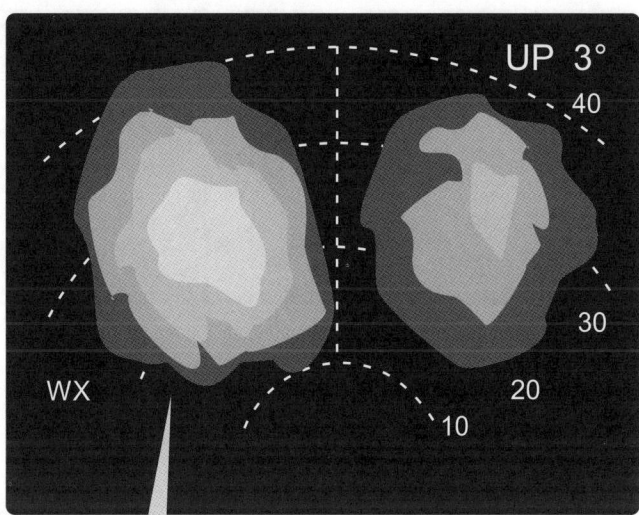

The cell on the left contains more abrupt changes in intensity levels, indicating the presence of severe turbulence.

**7-11: Sharp intensity gradients suggest severe turbulence.**

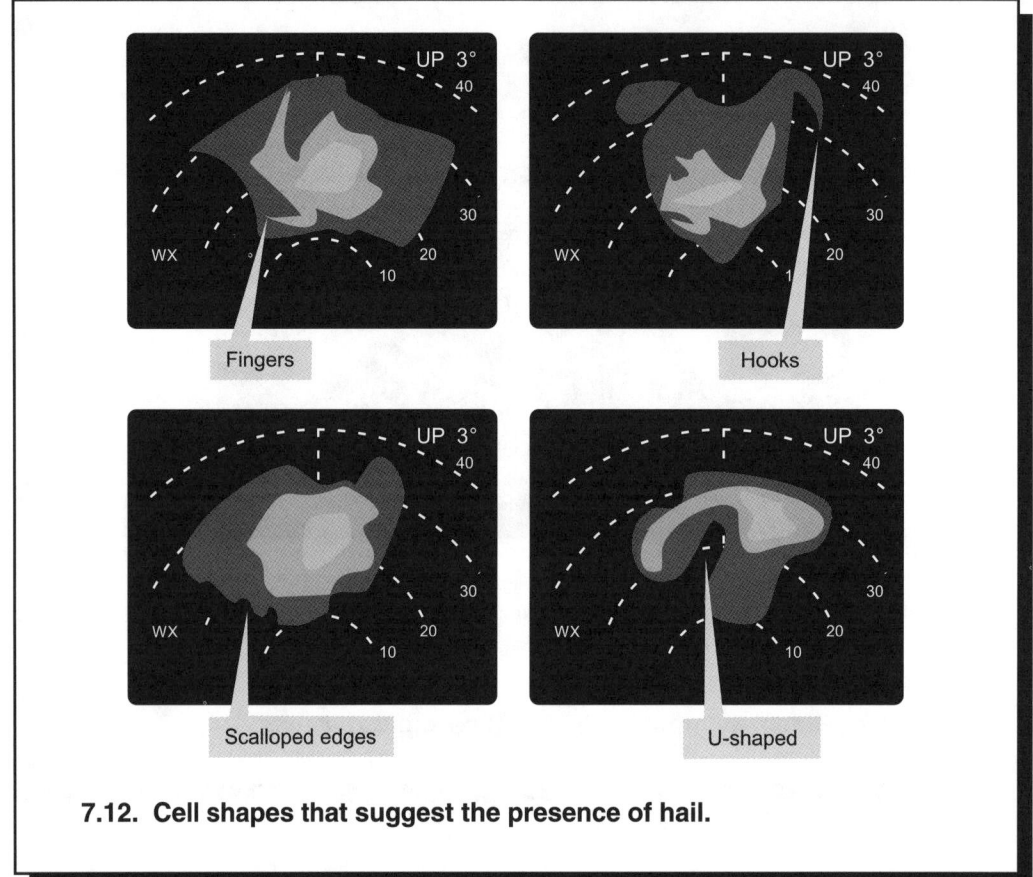

7.12. **Cell shapes that suggest the presence of hail.**

### *Adjusting the Antenna Tilt*

Every weather radar system relies on an **antenna** to transmit and receive radio signals. Radio signals emit from this antenna in a focused beam, similar to the way a flashlight emits light. Figure 7-13 illustrates how a weather radar system can only "see" a portion of what lies ahead of you at any one time.

In order to get a more complete picture of what lies ahead of the airplane, you must adjust the **tilt** of the radar system's antenna. This is accomplished using the **tilt control** located on the front of the radar display. This control allows you to adjust the tilt of the antenna 15 degrees up or down. The adjustment of the antenna tilt is an important part of being a proficient user of weather radar.

To see why antenna tilt adjustment is so important, let's start by adjusting the antenna tilt all the way down as shown in Figure 7-14. With the antenna set this low, many of the echoes are reflections from the ground. Additionally, the radar now fails to see the weather phenomenon above the terrain.

# 7 / Avoiding Traffic, Terrain, and Weather 143

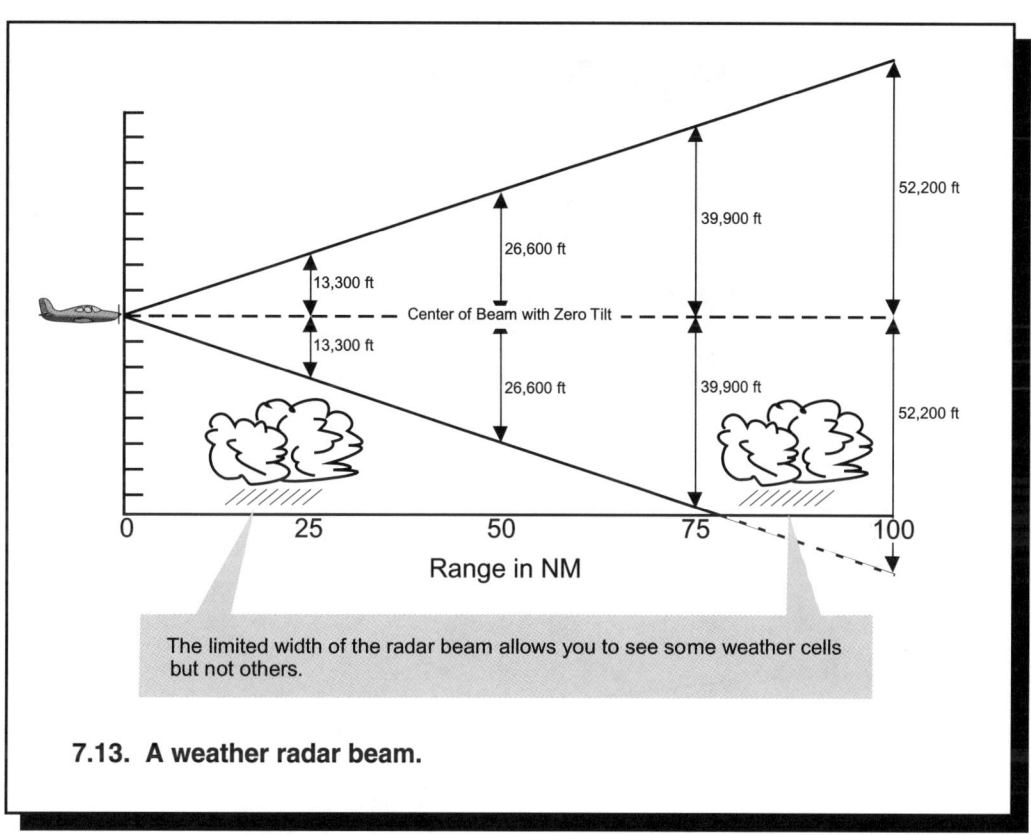

The limited width of the radar beam allows you to see some weather cells but not others.

**7.13. A weather radar beam.**

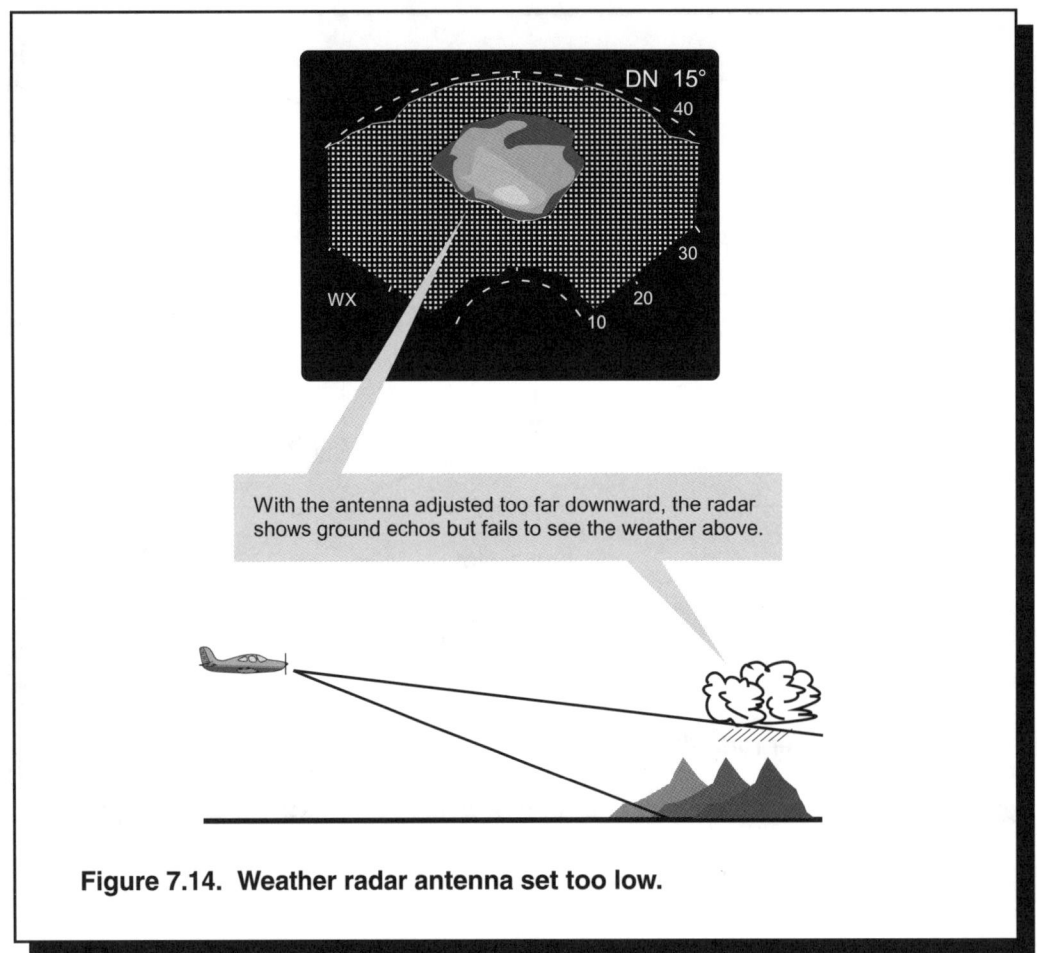

**Figure 7.14. Weather radar antenna set too low.**

Now let's set the antenna tilt too high, as shown in Figure 7-15. In this scenario, the radar fails to see the worst part of a significant thunderstorm cell located in front of the airplane.

These two situations illustrate how water masses can be missed during straight-and-level flight when antenna tilt is not used strategically. Think about how much worse the problem gets during a climb or a descent, when the airplane itself is tilted, so to speak. As you can now see, there's more to operating a weather radar system than flipping a switch and watching a video screen.

### Diverting Around Weather

The very purpose of an airborne weather radar system is to allow you to avoid inadvertent flight into thunderstorms and their associated hazards. There are some general rules for reacting to thunderstorm cells that appear on your radar. A first rule is to recognize upcoming hazards and make plans to fly away from them *as early as possible*.

# 7 / Avoiding Traffic, Terrain, and Weather

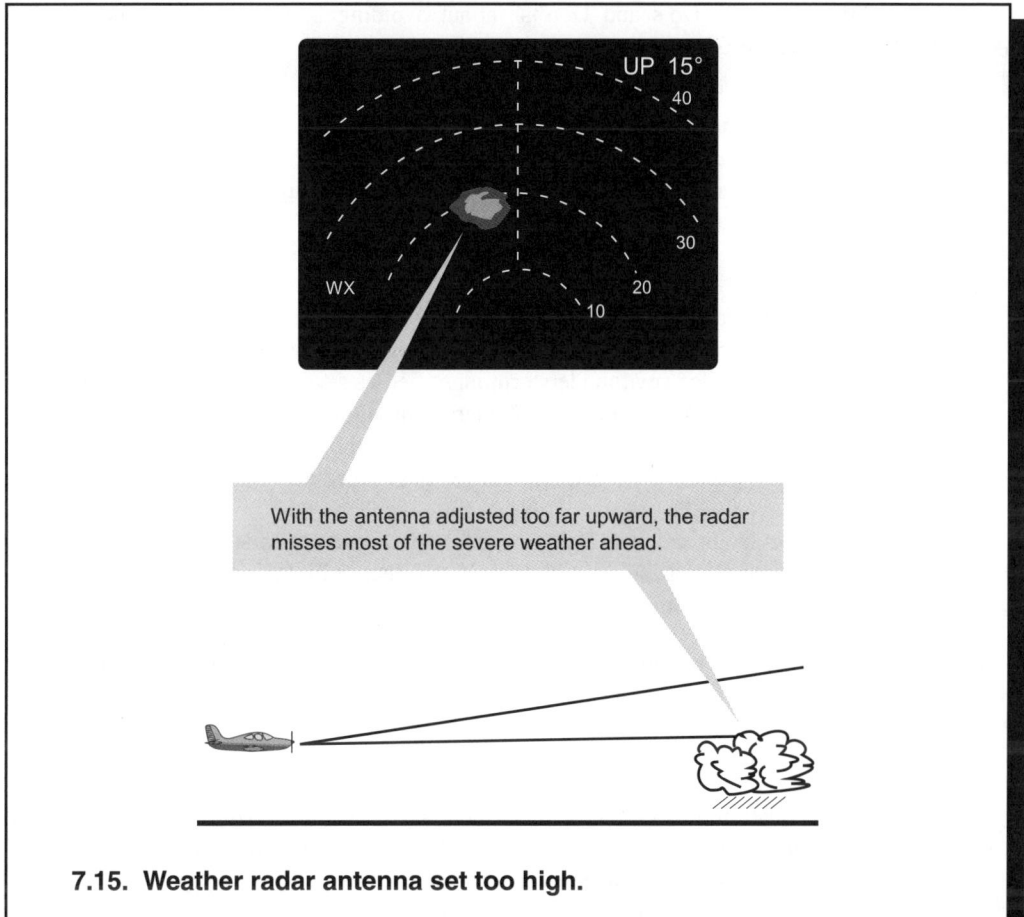

7.15. Weather radar antenna set too high.

The FAA offers several "Do's and Don'ts" about avoiding thunderstorms in its *Aviation Weather* publication (AC 00-6A).

- Don't land or take off in the face of an approaching thunderstorm. A sudden wind shift or low-level turbulence could cause loss of control.
- Don't attempt to fly under a thunderstorm even if you can see through to the other side. Turbulence under the storm could be disastrous.
- Don't try to circumnavigate thunderstorms covering 6/10 of an area or more either visually or by airborne radar.
- Don't fly without airborne radar into a cloud mass containing scattered embedded thunderstorms. Scattered thunderstorms not embedded can be visually circumnavigated.
- Do avoid by at least 20 miles any thunderstorm identified as severe or giving an intense radar echo. This is especially true under the anvil of a large cumulonimbus.
- Do clear the top of a known or suspected severe thunderstorm by at least 1,000 feet altitude for each 10 knots of wind speed at the cloud top. This would exceed the altitude capability of most aircraft.
- Do remember that vivid and frequent lightning indicates a severe thunderstorm.
- Do regard as severe any thunderstorm with tops 35,000 feet or higher whether the top is visually sighted or determined by radar.

Other experts additionally recommend choosing the upwind side of a thunderstorm cell for a diversion path rather than the downwind side. This helps ensure that the thunderstorm will not drift into your flight path.

Other do's and don'ts of weather radar include the following:

- Don't use the information provided by weather radar as a reason for taking more risk than you normally would without the weather information. Weather radar was designed to help you fly away from thunderstorms, not in, under, over, or between them.
- Be aware that the effectiveness of weather radar is determined by the skill of the operator. Learning to use antenna tilt, interpret radar images, and plan diversions are acquired skills. Don't be led into a false sense of security just because you have a sophisticated piece of equipment onboard your airplane. The equipment will only be as effective as its operator.
- A weather radar system is no substitute for a good weather briefing followed by in-flight updates.
- Know the many limitations of weather radar. For example, weather radar is not able to detect most other kinds of hazardous weather such as fog, icing, and turbulence.
- When a cell is detected by weather radar, that cell often absorbs or reflects all of the radio signals sent out by the radar system. This prevents the radar from detecting any additional cells that might lie behind the first cell.

Figure 7-16 illustrates this phenomenon known as **radar attenuation** in which one cell "shadows" another cell. The rule is to never attempt to fly into the shadow of a radar echo: You never know what might lie in there.

# 7 / Avoiding Traffic, Terrain, and Weather

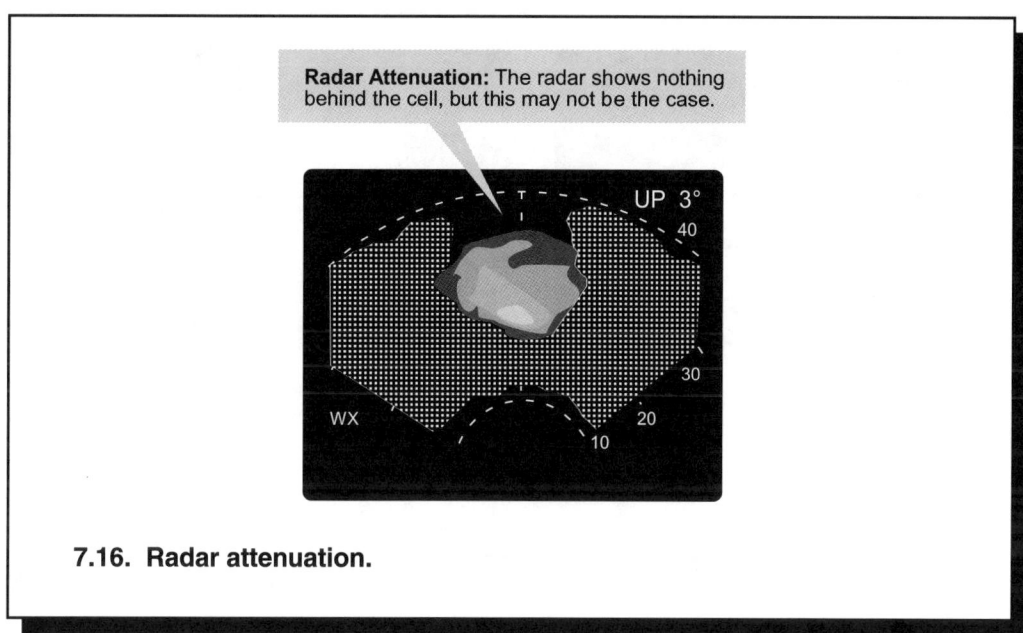

7.16. Radar attenuation.

Another limitation of weather radar is that the earliest stage of a thunderstorm is usually free of precipitation and may not be detected by radar. Despite the lack of precipitation, convective wind shear, severe turbulence, and icing are characteristic of thunderstorms during their cumulus stage.

Figure 7-17 offers a lesson for the operator on the importance of setting the range on the weather radar display. Looking at the display on the left, it would seem that you have a clear path between two thunderstorm cells. Changing the range to extend farther, you can now see the more massive cell that lies beyond them. Imagine flying between the two cells and seeing the third cell when it's already too late.

## Lightning Detection Systems

A second kind of system capable of detecting thunderstorms attempts to detect lighting that occurs within thunderstorm cells. A **lightning detection system** consists of a simple antenna and processing unit that senses electrical discharges in the atmosphere and attempts to determine which electromagnetic signals have the "signature" of lightning strikes. The displays for two popular lightning detection systems are shown in Figure 7-18.

Lightning detection systems are an excellent complement to a weather radar system. Both systems have their strengths and limitations and work nicely together to present a more complete weather picture. For this reason, many airplanes have both types of systems.

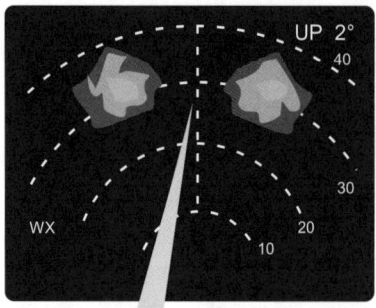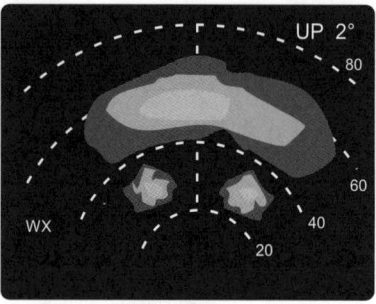

Different range settings can sometimes present very different pictures. Imagine flying through this gap ...

... only to find it wasn't much of a gap afterall.

**7.17. Importance of range settings on the weather radar display.**

Electrical discharges that are characteristic of lighting are plotted on these **lightning detection system displays.**

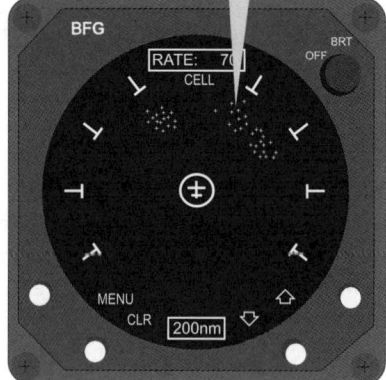

**7.18. Lightning detection system displays.**

# 7 / Avoiding Traffic, Terrain, and Weather

 **PRACTICE SESSION**

Unfortunately, at the time of this writing, there are few cost-effective ways of getting hands-on practice with the cockpit automation systems discussed in this chapter. Very few rental airplanes contain this kind of advanced cockpit automation. Several companies and individuals offer seminars and private instruction. For the pilot in training with limited funds, reading about these systems may be the only opportunity to learn while you are waiting to land that airline job.

# CHAPTER 8

# The Human Factor

*Throughout the chapters in this book, we have discussed potential problems that can arise when human pilots use cockpit automation systems. The problems that arise do not necessarily represent a flaw in the design of the automation or a defect in our piloting skills. Rather, they seem to be a phenomenon that occurs naturally when humans work with computers. It is no surprise that there is an entire academic discipline dedicated to the study of what are called human factors. In this chapter, we take a look at some of the important known human factors issues that experience with cockpit automation has brought to our attention.*

In this chapter, we will review a variety of issues that surround the use of cockpit automations such as the ones we have discussed throughout the book. Some of these issues were discussed in other chapters, but it is useful to revisit them now that you have acquired some automation expertise and are experiencing life in an automated cockpit.

## BECOMING AN "INTELLIGENT USER" OF COCKPIT AUTOMATION

If you get the impression that learning about cockpit automation is harder than learning about other kinds of aircraft systems, don't feel alone. The new generation of automated aircraft systems is different from systems you have used in the past. Cockpit automation systems are more difficult to understand than traditional electrical or mechanical systems. While mechanical control surfaces and radio navigation aids have moving parts that you can see and simple electrical circuits that you can follow, systems like GPS computers have software and logic that are much more complex. Furthermore, this logic is hidden from plain view and can only be understood through reading, listening to others, and experimenting. Cockpit automation systems are no less perplexing than the personal computers we use in our office or at home.

Can you ever truly master a system as complex as a GPS computer? Probably not. The software used in GPS computers was written by teams of computer programmers, with each programmer understanding how only a piece of the entire system works. In many cases, there may be no single living human being that completely understands how the entire system works, so don't waste time feeling stupid. So how do you do a good job of operating a system that you will never completely understand? The key is to commit yourself to being an "intelligent user" of cockpit automation.

An intelligent user of cockpit automation is one that recognizes the limits of their own understanding and acts accordingly. Two examples nicely illustrate this point. Recall from Chapter 5 how you can sometimes get into trouble using the nonsequencing mode when intercepting an approach course. When you engaged the sequencing mode prior to reaching the approach course, the computer mistakenly sequenced through some of your approach waypoints and sent you directly to the missed approach point. Even though you didn't understand why this happened, you know now that it is an important limitation of the automation and that you must act accordingly. Don't use any automation feature in any situation unless you are confident that it is going to work properly. Avoid any temptation to blindly trust cockpit automation just because "it's a computer." Let the automation earn your trust over time by demonstrating to you how it performs in specific situations.

Consider another case in which you are asked to fly a DME arc. The GPS computer manual indeed boasts a powerful capability for flying DME arcs, but you haven't learned about it yet. The manual claims that the GPS computer trivializes the problem of intercepting the arc at any point and automatically takes care of any needed adjustments to the desired track to the active waypoint. If you were instructed to fly a DME arc, would you want to just jump in and try this feature? Given that you already know that the computer can get you into trouble just as easily as it can get you out, don't try anything new until you've had a chance to learn more about it. A busy terminal area is no place to be fixating on a GPS computer in an attempt to learn something new. In this case, why not use conventional techniques for flying a DME arc or ask ATC for a different approach?

An intelligent user of automation commits to an ongoing learning process. Even though you know you won't ever understand everything about the automation, everything that you do know will probably help you at some point. Be as smart as you can.

An intelligent user of automation should expect surprises from the automation. These surprises can arise in situations of which neither the pilot nor the engineer is aware. Remember the three commonly asked questions in the automated cockpit: (1) Why did it do that? (2) What's it doing now? and (3) What's it going to do next? The best way to deal with surprises is to be prepared for them. In order to be prepared for surprises, you need to be paying close attention to what is going on. Experience with cockpit automation has shown that paying attention isn't always as easy as it looks, a topic that we turn to now.

## STAYING IN THE LOOP

An inevitable problem faced by users of cockpit automation is the problem of remaining aware of what is going on around you while the automation does its work. During your practice sessions, you probably got caught forgetting to check whether or not the sequencing, nonsequencing, terminal, or approach modes were engaged. If someone were to cover up the navigation page on the GPS computer, would you always be able to tell where you were?

It is well documented that using automation challenges our ability to "stay in the loop." But a degraded awareness of what is going on is not necessarily a product of pilots becoming lazy or complacent. Rather, our degraded awareness is more a natural feature of human attention. A basic limitation of human vision, audition, and cognition is that we can only look at, listen to, or think about one thing at a time. How do we succeed at tasks such as flying, in which many things are going on at once? The way humans solve this problem is to switch their attention back and forth between the things that seem the most important to them at any given time. Now, if our IFR GPS does an impeccable job of making its way from one waypoint to the next ninety-nine percent of the time, wouldn't it be a better idea to concentrate on other aspects of our flight. Are we picking up ice? Are all aircraft systems functioning properly? Have we sufficiently briefed our approach? In fact, the only time we would really want to pay lots of attention to the GPS is when something is going wrong, and herein lies the paradox of automation. The very reason cockpit automation was designed and built was to free pilots from the tedious task of continuously scanning instruments and making corrections and to attend to other matters. But the only way to ensure that the automation is working correctly is to "watch over the shoulder" of the automation and make sure the job is being done right. Now you have two jobs: your original job of flying the airplane and operating the automation. Indeed, airline pilots often remark that sometimes automation lowers your workload, and sometimes it raises it.

So what is the solution to the paradox of automation? The key is to realize that when automation is being used, you assume the role of a manager rather than a worker. You now have to look at what is going on through managerial eyes. As you learned when you were starting out flying instruments: Don't stare at anything for too long. Keep your attention moving. Give ample attention to your important matters-of-the-minute but also find time to check up on tasks that are being performed by the automation.

What do we mean specifically by the phrase *staying in the loop*? Suppose you are following a GPS course in sequencing mode and have engaged your autopilot after reaching your assigned cruising altitude. Aside from whatever duties you are attending to, the following mental scan illustrates the idea of staying in the loop.

### 1. Keeping the Right Attitude

Have a look at the primary instruments. Is the autopilot keeping the aircraft upright, on the desired heading or track, and at the assigned altitude?

### 2. What Targets Is the Automation Trying to Achieve?

Look at the GPS computer and see where you are going next. What's the active waypoint that you are working toward? How far are you from the waypoint? What is your ground speed and when can you expect to be there? What's the next waypoint after the active waypoint and how big of a turn will you be making? Is your airspeed close enough to what you filed in your flight plan?

Awareness of targets is even more important during en route climbs and descents and during terminal procedures. This is because your targets are changing more often. Did you dial in the right inbound course when you switched to OBS mode?

### 3. What Modes Are Currently Engaged?

Are you operating in the sequencing or nonsequencing mode? Keeping track of this is especially important for maneuvers in which you have to remind yourself to switch from one mode to the other at some critical time. Humans are particularly bad at remembering things they have to do in the future. In fact, psychologists have made an entire academic discipline (called **prospective memory**) out of studying how people remember "things-I-really-need-to-remember-to-do." Throughout history, people have invented ways of improving their performance at prospective memory. Tying a string around your finger, notes to self, timers, alarms, and checklists are all time-honored techniques. How are you going to remember to check for the approach mode on your GPS computer when you reach the final approach fix?

## SHARING YOUR INTENTIONS, GOALS, AND INFORMATION

If you have plans to go on to a professional aviation career, you will likely find yourself in an automation-equipped aircraft operated by more than one pilot crewmember. In this situation, you now have yet another resource available for accomplishing the duties of the flight. This situation presents opportunities for effective teamwork but also some interesting possibilities for confusion. Suppose you and your copilot agree that the copilot will fly the airplane while you handle the radio communications and make entries into the GPS computer and autopilot. This is an excellent strategy since it allows the copilot to keep his or her eyes on the instruments and out the window whenever "heads-down" programming needs to be done on the GPS computer. Now suppose ATC issues radar vectors to the final approach course. You read back the clearance, select OBS mode, and twist in the final approach course. How does your copilot know what you just did and whether or not you did it right? What if you twisted in the wrong approach course? Would your copilot blindly follow the needle across the true approach course?

When automation is used in a crew environment, the crew must develop effective communication strategies—procedures for ensuring that each crewmember and the automation has received the same message and has the same plan in mind. The airlines use established procedures for cockpit automation use that work well. For example, when the controller issues the radar vector to the final approach course, you can read back the clearance and then announce that you are selecting OBS mode and twisting the final approach course. Your copilot can then take a brief moment to check your work. But how do you know that your copilot indeed checked what you did? To solve this problem, your copilot can use a similar procedure: Point at the GPS computer and say "OBS mode: Here's where we're going and here's how we're getting there." Procedures such as these help ensure that all parties have the same plan in mind.

## BLUNDERS MADE EASY

When using cockpit automation, small actions sometimes have big consequences. Throughout our discussions, you saw several examples of how missed or mistaken steps could lead to serious consequences. For example, forgetting to set the GPS computer back to the sequencing mode during a vectored approach can, in some cases, lead you to fly through an approach course. This is a fairly severe penalty for forgetting to push one button. Mistyping the name of a waypoint is another simple mistake that could potentially cause you to turn your aircraft off the intended course.

Beginning with the idea that "to err is human," an important goal in designing cockpit automation is to make systems that are more forgiving of human error. We call this kind of system an error-tolerant system—one that attempts to make mistakes evident to the human operator and to limit the consequences of subtle mistakes. Your GPS computer indeed offers several error-tolerant features that help limit the occurrence and consequences of errors. The map display that appears on the navigation page helps you to detect mistaken or misspelled waypoint names. For example, if you want to enter the Las Vegas VOR located in Nevada (LAS) into your flight route and mistakenly enter the Las Vegas VOR located in New Mexico (LVS), the map display will show your course veering off in an odd direction, as shown in Figure 8-1.

Instead of seeing the LAS VOR northeast of Palmdale, you see your course extending indefinitely in an unexpected direction. Looking at the map should alert you to the possibility that something is not right.

Another error-tolerant feature of the GPS computer is to require menu selections for terminal area procedures. By choosing an approach from a prepackaged list of approaches, the GPS unit relieves you from having to type in the individual approach waypoints. Imagine the consequences of mistyping a missed approach waypoint and taking a wrong turn as you descend down to the MDA!

## LOSS OF MANUAL FLYING SKILLS

In a popular survey of airline pilots, fifty percent of all pilots expressed concern about losing their manual flying skills due to excessive use of cockpit automation. Ninety percent of the pilots reported that they manually operated the aircraft during a portion of the flight in order to keep their skills sharp. Use this information to your benefit and don't let the computer always do everything. It is particularly tempting to become reliant on a GPS computer for position finding and tracking. The

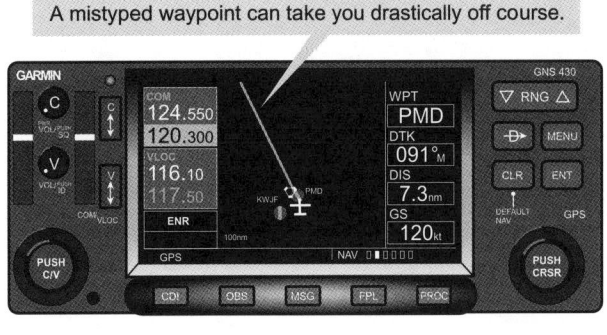

8.1. A small error can sometimes have big consequences.

visualization skills that you were taught throughout your training will never become obsolete. Don't get caught being rusty on the one day that your computer doesn't work.

## LOSS OF AUTOMATION SKILLS

The same potential for skill loss applies to automation use. During the development and testing of this book, I noticed that pilots who took time off between flights required more time to refresh themselves on the concepts and procedures they had mastered during previous flights. Cockpit automation often requires you to memorize procedures for which there are few visible cues or reminders once you step into the cockpit. If you plan on using cockpit automation under IFR, plan to spend time maintaining "automation currency" in addition to your conventional IFR currency.

# CHAPTER 9

# Proficiency Standards

*In this chapter, you get to serve as your own designated pilot examiner (a rare opportunity indeed). This exercise is your chance to determine whether or not you have attained proficiency in all of the areas covered in the book. Use the list of skills below as a checklist and make sure you are ready to demonstrate each skill as if you were going on a check ride. You can use the simulators to run through each of the skills that use the GPS computer. If you have the opportunity to get in an airplane, plan a flight that allows you to demonstrate at least one approach via vectors to the final approach course followed by a missed approach and hold procedure.*

## SKILLS CHECKLIST

SKILL 1: Determine if your GPS computer is approved for IFR navigation.
SKILL 2: Program the GPS computer with a flight plan.
SKILL 3: Check the flight route to ensure that you and the computer have the same plan.
SKILL 4: Ensure adequate GPS signal reception.
SKILL 5: Demonstrate how to follow and monitor your progress along the route stored in the GPS computer.
SKILL 6: Build and fly a descent.
SKILL 7: Demonstrate a GPS approach explaining all approach mode transitions.
SKILL 8: Demonstrate the Direct-To function.
SKILL 9: Add and delete waypoints from your flight route.
SKILL 10: Select a different approach or transition.

SKILL 11: Demonstrate an emergency diversion.
SKILL 12: Intercept an inbound course to the active waypoint.
SKILL 13: Intercept an inbound course to a different active waypoint.
SKILL 14: Holds.
SKILL 15: Procedure turns.
SKILL 16: Missed approaches.
SKILL 17: Explain the importance of continually monitoring the GPS computer.

## HOW DID YOU DO?

Congratulations! You have made it through the entire cockpit automation course. How do you think you did? Do you feel comfortable enough with the equipment to file as a "/G," flip on the autopilot, and go grind up some clouds? Before you answer in the affirmative, think about the ways in which the GPS computer and autopilot got the best of you during your flights. Do you have procedures that ensure that you'll never forget to switch back to the sequencing mode once you intercept the approach course? Are you always checking for the approach mode before descending to the MDA? Do you always know what autopilot functions are armed and engaged? If your answers are less than definitive, there is nothing wrong with getting some more practice and experience. File as a "/G" under day VFR conditions and practice going through the motions a few more times. As you move on and learn about more complex cockpit automation systems, you will quickly realize that dealing with cockpit automation is a lifelong learning experience. You will discover new ways to do things everyday and learn new ways in which the automation can surprise you. Consider yourself the proud owner of a "license to learn" and use your best judgment at all times. Learn all you can about automation and use it when you are confident that you are flying it and not the other way around. For now, welcome to the exciting world of cockpit automation and enjoy the ride.

# CHAPTER 10

# For Airline-Bound Pilots

*If you are a pilot with aspirations to move into an airline job, this chapter introduces you to the kind of cockpit automation systems you will encounter in the modern airline cockpit. You will see that if you have taken the time to master the skills presented in this book, you have ideally prepared yourself to make the transition to more sophisticated cockpit automation.*

As the book's title suggests, the set of skills you have learned here really has two purposes. For the general aviator, you are now on your way to being a proficient and safe operator in the new high-tech small-airplane cockpit. Your next step is to go out and get the flying experience needed to put your new knowledge and skills in motion.

For the aspiring airline pilot who uses small airplanes mainly as a training platform, the goal of this book was to provide you with the core skills you will need to transition to the more sophisticated equipment found in jet transport airplanes. In this chapter, you will see how everything you have learned in this book has ideally prepared you to meet the challenge of the modern airline cockpit.

## MOVING ON TO THE "BIG BOOK"

If you are an aspiring airline pilot and have gotten some experience using GPS and autopilots, your next step is to tackle the "big book." *The Pilot's Guide to the Modern Airline Cockpit* (see Figure 10-1) takes you through the same program of topics, skills, and concepts that you have mastered here. The difference is that, instead of working in a small training airplane, you will now be strapped into the seat of the next-generation Boeing 737, 747-400, and the Canadair Regional Jet.

You will learn how the GPS computer, autopilot, and other computers you have become familiar with are nothing less than miniature versions of the equipment found in the modern airline cockpit. You will find that the knowledge and skills you have acquired here have ideally prepared you to meet that final frontier of cockpit automation.

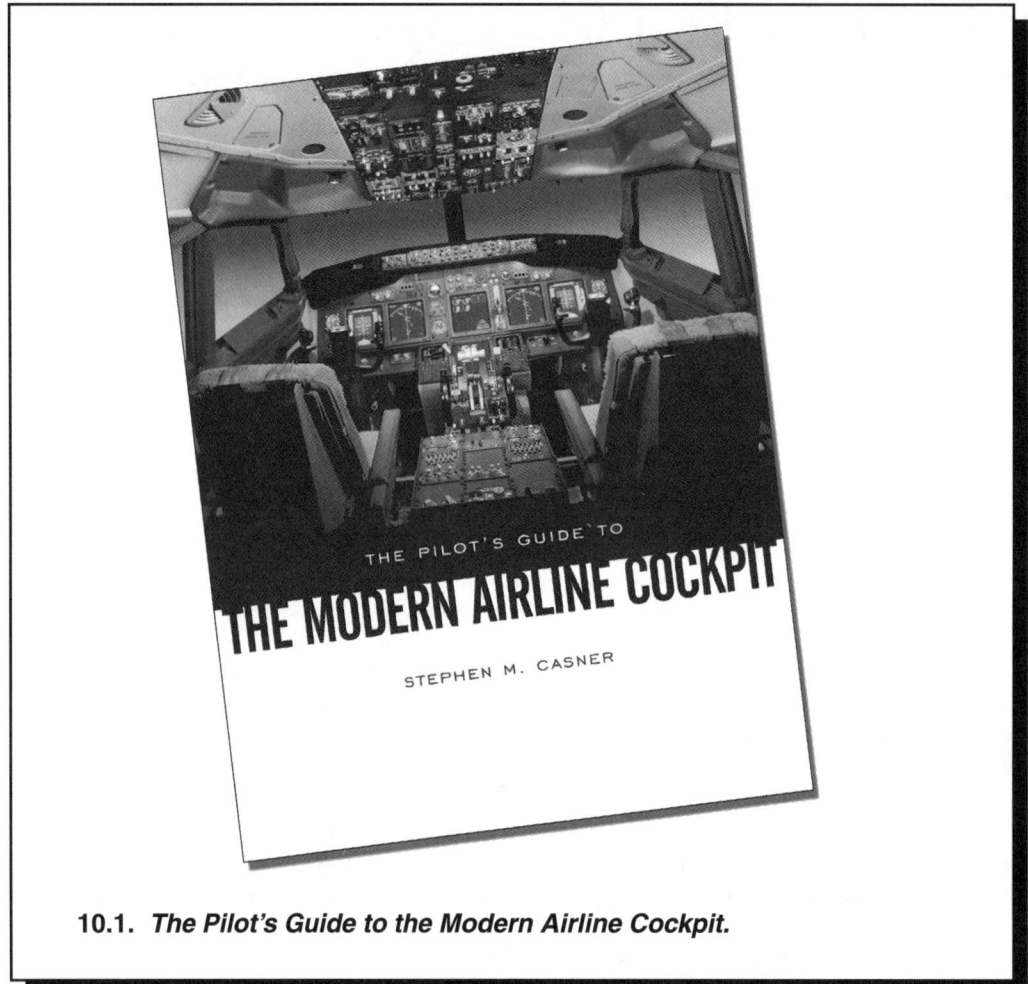

**10.1.** *The Pilot's Guide to the Modern Airline Cockpit.*

Airline carriers have struggled with cockpit automation training since the introduction of these systems in the early 1980s. Walking into a job interview with a good working knowledge of these systems is going to look impressive.

Let's take a quick tour through the next-generation Boeing 737 cockpit and see how prepared you will be to make this transition.

## The Flight Management Computer

The **flight management computer** (**FMC**) found in every modern airline cockpit is really nothing other than a more sophisticated version of the GPS computer that you are already familiar with. Let's look at some of the differences between the little computer and big computer.

You'll be pleased to learn that your days of twisting knobs to make entries are over. The flight management computer has a separate and much larger **control display unit** (**CDU**) that lets you look at pages and type in characters. Shown in Figure 10-2, the control display unit offers a larger screen you can use to view pages, along with a keypad for making entries.

# 10 / For Airline-Bound Pilots

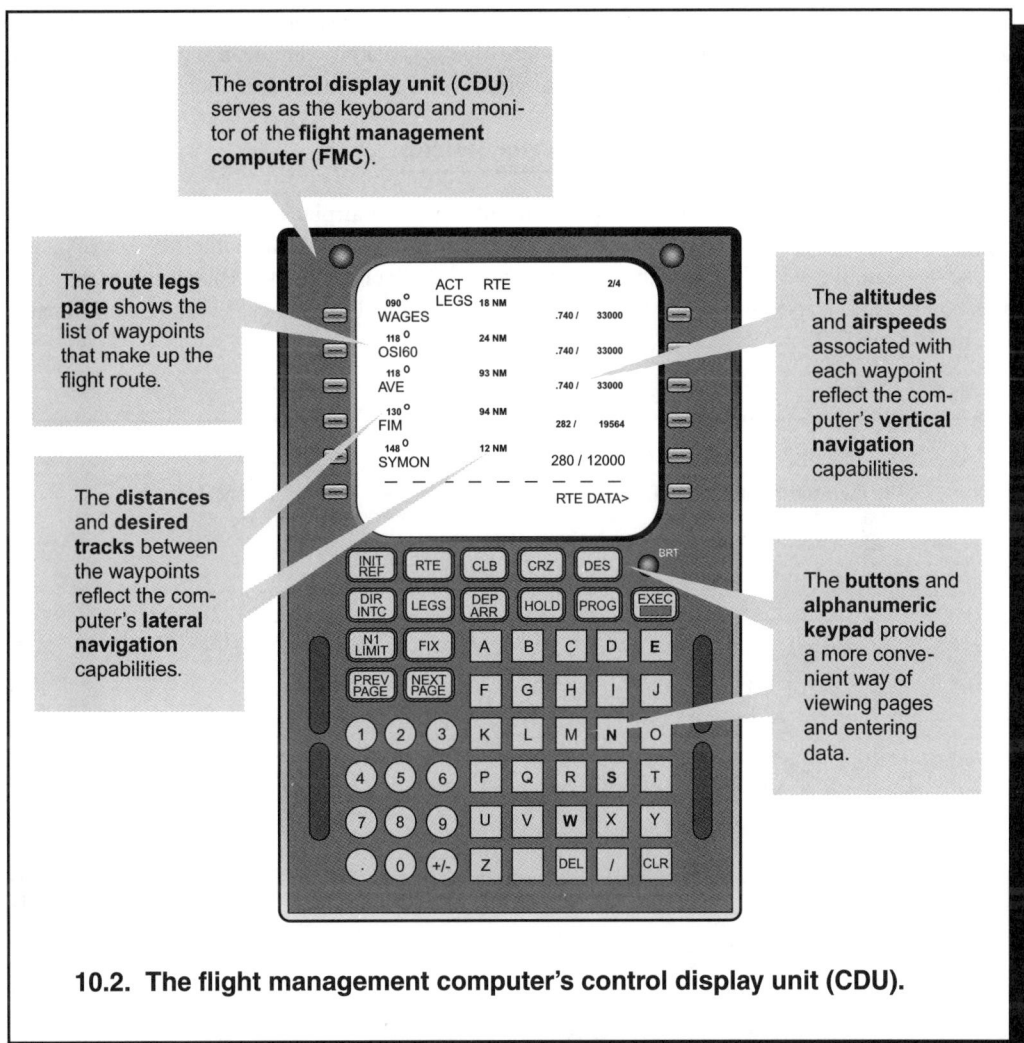

**10.2. The flight management computer's control display unit (CDU).**

The CDU in Figure 10-2 shows the **route legs page** on the control display unit. The route legs page is similar to the flight plan page on your GPS computer. The route legs page also lists the sequence of waypoints that make up your route. Just like your GPS computer, notice that the flight management computer figures out the distances and desired tracks between the waypoints.

Looking at the right side of the route legs page, notice that the flight management computer shows a planned airspeed and altitude for each waypoint in your route. This is perhaps the biggest difference between your GPS computer and a flight management computer. Not only does a flight management computer help you navigate between the waypoints in your route (what we call **lateral navigation**) but also up to and down from your planned cruising altitude, obeying any and all crossing restrictions along the way. This **vertical navigation** capability of the flight management computer even extends to helping you figure out what thrust settings should be used during your climbs, cruise, and descents.

Similar to the navigation page on your GPS computer that presents a graphical depiction of your position and planned flight route, the modern Boeing cockpit contains a separate **navigation display** (**ND**) that shows this same information. Figure 10-3 shows the navigation display.

## The Autopilot, Autothrottle, and Flight Director Systems

Aside from computers that help you plan a flight route, big airplanes also have **autopilots** that help you steer the airplane along the route you have planned. Similar to the autopilot control panels you are now familiar with, big airplanes feature a **mode control panel** (**MCP**) that offers the same collection of buttons used to engage autopilot functions. Figure 10-4 shows the mode control panel from the Boeing 737.

The **navigation display** (**ND**) shows a graphical depiction of the present position of the airplane and the route stored in the flight management computer (FMC).

10.3. The navigation display (ND).

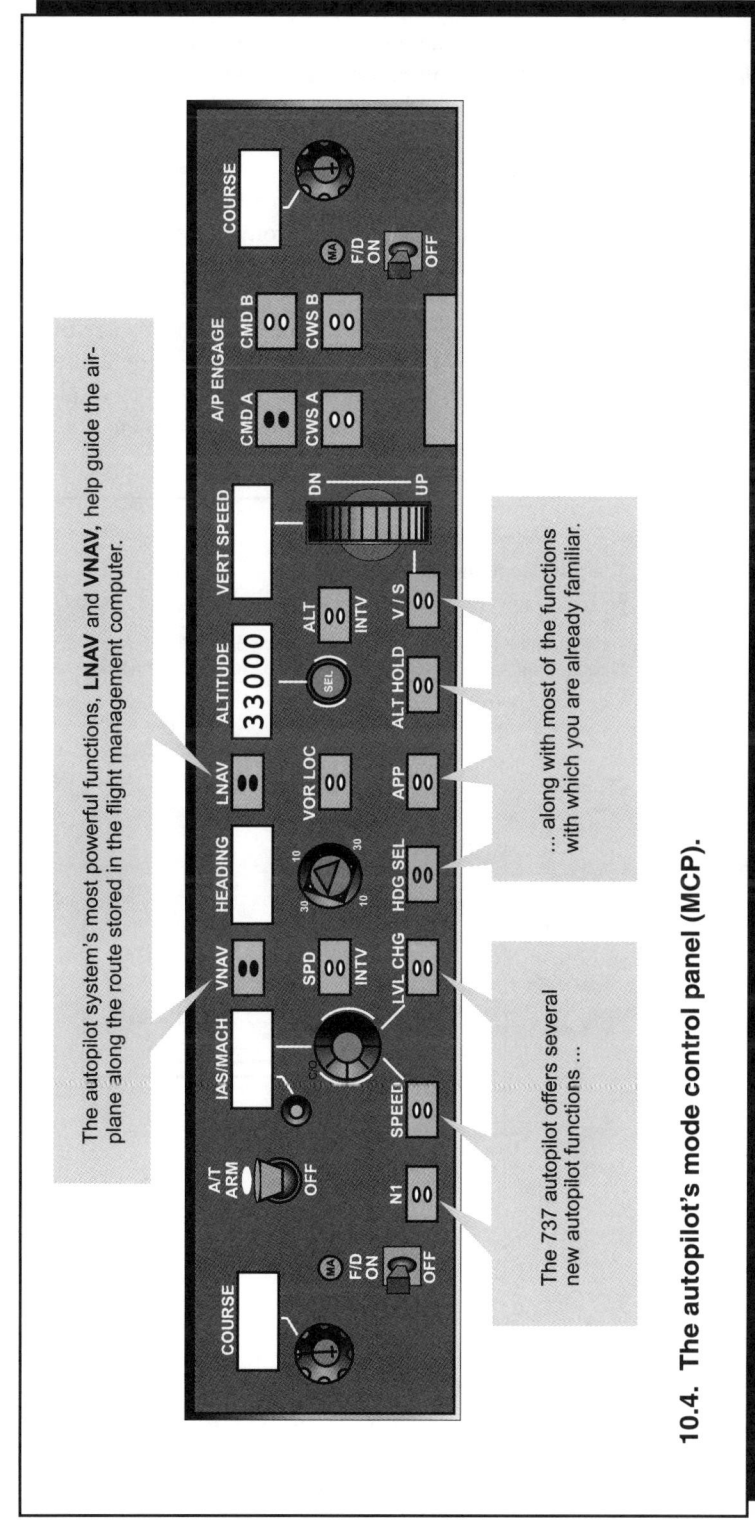

10.4. The autopilot's mode control panel (MCP).

As you can see, the autopilots in big airplanes offer you a few more functions. Like the navigation function found in little airplanes, big autopilots offer you a function called **lateral navigation (LNAV)**. LNAV helps guide you between the waypoints in your programmed route (shown on the route legs page). Unlike little airplanes, the big ones offer you a second function called **vertical navigation (VNAV)**. VNAV helps guide you to the target altitudes and airspeeds that appear beside each waypoint on the route legs page. The flight management computer calculates the thrust settings needed to achieve these vertical targets. Big airplanes also feature an **autothrottle system.** The autothrottle system is capable of automatically controlling the airplane's thrust levers. When the flight management computer decides it is time to climb or descend, it instructs the autothrottle system to make the appropriate thrust adjustments.

How do you stay in the loop while all this automation works its magic? Big airplanes like the Boeing 737 also feature a **flight director.** Figure 10-5 shows a set of flight director command bars presented on the airplane's **primary flight display (PFD)**. This all-in-one display presents attitude, altitude, airspeed, vertical speed, the flight director, and much more, in one combined package.

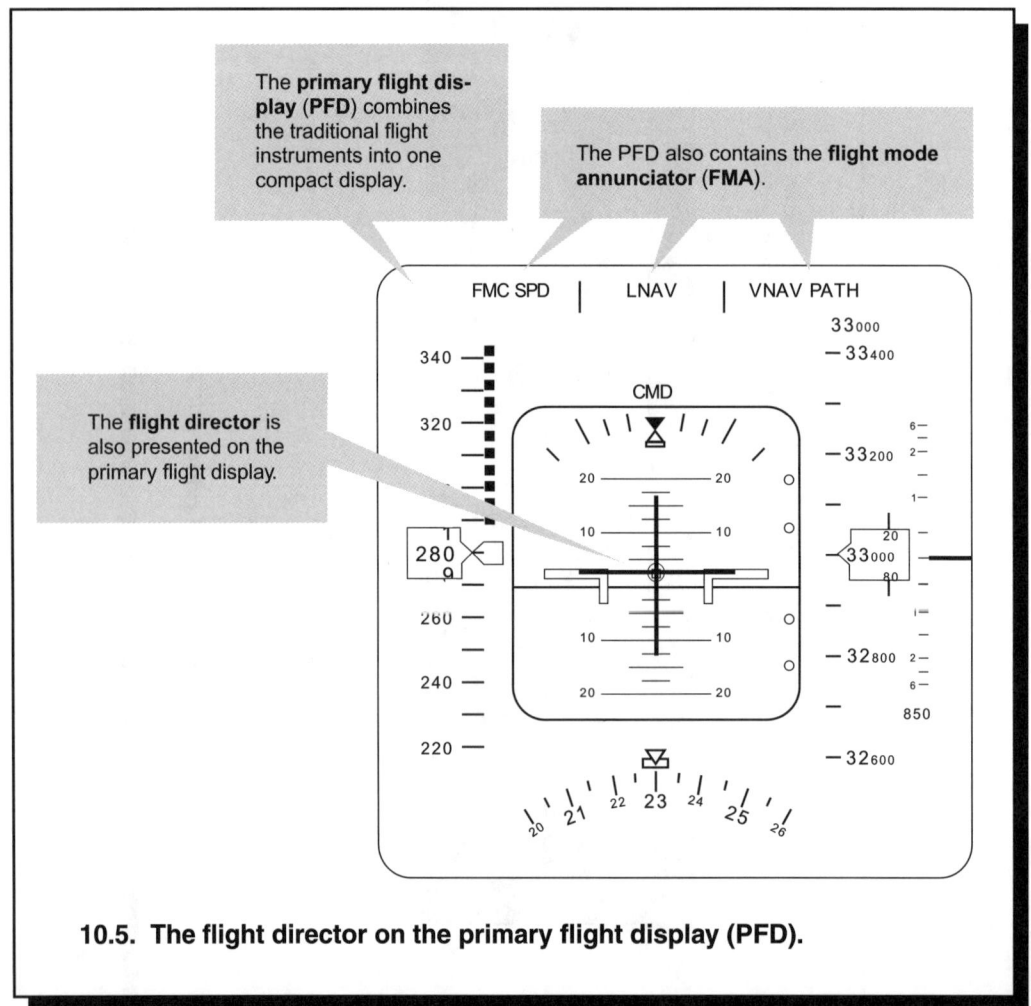

**10.5. The flight director on the primary flight display (PFD).**

The top of the primary flight display contains the **flight mode annunciator** (**FMA**). Of course, you have not forgotten that the flight mode annunciator is where you must regularly check to stay on top of what autopilot and autothrottle functions are currently armed or engaged. Study this FMA (Figure 10-5) more closely: Big airplanes have more functions to keep track of.

## TCAS, GPWS, and Weather Detection Systems

If you have taken the time to learn about the traffic collision avoidance system (TCAS), ground proximity warning system (GPWS), and weather detection systems available in small planes, I have good news for you. The pilot interfaces to these systems are mostly standardized, and you already know almost everything you need to know about how they work. The only thing you'll need to add to your bag of tricks now are the standard operating procedures used by your airline.

# GLOSSARY

# Terms and Abbreviations

**Airborne weather radar:** An onboard system capable of detecting significant masses of precipitation. The primary use of weather radar is to aid the pilot in avoiding thunderstorms and their associated hazards. It is a complex system, and skill and practice are required to become a proficient user of weather radar.

**Altitude alerter:** The system that allows you to dial a target altitude and then receive a visual and/or auditory alert when the airplane reaches the vicinity of the dialed altitude. Designed as a backup to pilot altitude awareness.

**Altitude capture:** An autopilot function that enables the autopilot to automatically level the airplane at a selected altitude.

**Altitude function:** An autopilot function that allows you to capture and maintain the present altitude of the airplane.

**Altitude selector:** The cockpit control and display that allows you to dial an assigned altitude and instruct the autopilot to climb or descend to that altitude.

**Approach function:** An autopilot function that allows you to capture and track any GPS course, VOR radial, or localizer with a higher degree of accuracy.

**Autopilot:** The system that automatically manipulates the roll and pitch of the airplane to follow the route programmed into the GPS unit or to follow simpler altitudes, vertical speeds, headings, and courses dialed by the pilot.

**Autopilot control panel:** The device used to engage autopilot functions. Usually combined with the autopilot flight mode annunciator.

**Autopilot master switch:** The switch that turns on the autopilot and flight director systems. No autopilot or flight director functions are available unless the autopilot master switch is on.

**Autothrottle:** A system found in larger jet transport airplanes that automatically manipulates the thrust setting of the airplane to help follow the vertical trajectory portion of the planned flight route.

**Autotrim:** The system that automatically adjusts the pitch trim of the airplane in response to trim commands generated by the autopilot.

**Bottom-of-descent point:** The endpoint of the descent calculated by the GPS computer. The bottom-of-descent point typically corresponds to a descent crossing restriction that you have been asked to meet.

**Cockpit automation:** The term that loosely refers to all of the electronic systems in the modern cockpit designed to support the pilot in planning and carrying out a flight route.

**Command bars:** Symbols that appear on the flight director to indicate the roll and pitch commands that are generated by the autopilot.

**Cursor mode:** The function offered by the GPS computer that allows you to input data to the computer. Data inputs are made using the same knobs used to switch between pages.

**Desired track:** The great circle course computed by the GPS computer that takes you from the previous waypoint in your flight plan to the active waypoint.

**EFIS:** *See* Electronic flight instrument system.

**Electronic flight instrument system:** The computers and displays that replace the traditional primary flight instruments. EFIS displays can also present information from navigation instruments, collecting much of the important information into one easily scanned display.

**Flight director:** A display that presents roll and pitch commands generated by the autopilot. Similar to following localizer and glide slope needles, the pilot flying must keep the airplane symbol lined with the command bars on the flight director.

**Flight mode annunciator:** A display that presents the names of the autopilot functions that are either armed or engaged. The flight mode annunciator is the only reliable source of information about what autopilot functions are in use.

**Global positioning system:** A satellite-based navigation system that can be used to determine the position and track the movement of an aircraft. A GPS computer must be installed onboard the aircraft to receive and interpret signals from the satellite-based system.

**GPS:** *See* Global positioning system.

**GPS computer:** A component that must be installed onboard the aircraft in order to make use of the global positioning system. This component receives satellite signals and determines and tracks the position of the aircraft.

**GPWS:** *See* Ground proximity warning system.

**Great circle route:** The shortest distance between two points when traveling on the surface of Earth. Defined by a plane that passes through the two points of interest and the center of Earth.

**Ground proximity warning system:** The system that provides visual and auditory alerts whenever the aircraft is determined to be in danger of colliding with surrounding terrain.

**Integrated TCAS vertical speed indicator:** A cockpit display that combines the traditional vertical speed indicator (VSI) with a traffic collision and avoidance system (TCAS) display.

**Leg mode:** The name for the sequencing mode on the Bendix/King GPS units.

**Lightning detection system:** An onboard weather detection system that senses electrical discharges that suggest the presence of thunderstorm cells.

**MFD:** *See* Multifunction display.

**Multifunction display:** A cockpit display capable of presenting different kinds of information.

**Navigation function:** An autopilot function that allows you to track the route programmed in the GPS computer or a VOR radial.

**Nonsequencing mode:** The GPS computer mode that does not automatically sequence between the waypoints in the programmed route. The nonsequencing mode maintains the current active waypoint indefinitely and allows you to specify your own desired track to that waypoint. Useful for course intercept maneuvers and holding patterns.

**OBS mode:** The name for the nonsequencing mode on Bendix/King and Garmin GPS units.

**Overlay approach:** A conventional approach procedure that relies on radio navigation equipment but that also has juxtaposed on top of it another approach procedure that uses modern area navigation equipment.

**Page:** Any one of a collection of information displays that can appear on the GPS computer. Every page has a title and presents information related to a particular navigation topic (e.g., airport elevation, runways, communications frequencies).

**RA:** *See* Resolution advisory.

**Radar attenuation:** The phenomenon in which a weather cell absorbs the radar pulses that are sent out by an airplane's weather radar system. The problem posed by radar attenuation for the pilot is that thunderstorm cells that lie behind another cell often go undetected since the radar pulses are absorbed by the first cell.

**RAIM:** *See* Receiver autonomous integrity monitoring.

**Receiver autonomous integrity monitoring:** The self-monitoring function performed by the GPS receiver unit to ensure that adequate GPS signals are being received at all times. The GPS computer will issue an alert to the pilot whenever the integrity monitoring determines that the GPS signals do not meet the criteria for safe navigation use.

**Resolution advisory:** A more serious warning issued by the traffic collision avoidance system (TCAS) indicating an immediate threat of collision with another aircraft. This warning takes the form of an aural command to perform a vertical avoidance maneuver (e.g., "Climb! Climb!").

**Sequencing mode:** The GPS computer mode that automatically sequences between the waypoints in the programmed route. The sequencing mode alerts you to upcoming waypoints and guides you through the turn to each successive waypoint in the route.

**Stand-alone approach:** An instrument approach that relies solely on the use of area navigation equipment such as a GPS receiver.

**TA:** *See* Traffic advisory.

**TCAS:** *See* Traffic collision avoidance system.

**TCAS control panel:** The cockpit device that allows you to select the different features of your traffic collision and avoidance system (TCAS).

**Terrain display:** A pictorial display that shows surrounding terrain and obstacles that present a potential threat to your aircraft, given your present altitude. Draws terrain information from a terrain and obstacle database.

**Terrain inhibit switch:** A switch that allows you to suppress all visual and auditory warnings given by the ground proximity warning system (GPWS). Often used to silence "nuisance alerts" when deliberately operating in the vicinity of terrain.

**Terrain and obstacle database:** An electronic database stored in your ground proximity warning system (GPWS) that details all of the significant terrain features and obstacles that could potentially pose a threat to aircraft flight. These data can be presented on a terrain display.

**Top-of-descent point:** The point that the GPS computer calculates to be the ideal location at which to begin a descent down to the planned crossing restriction, given the descent speed and rate that has been entered.

**Traffic advisory:** A warning issued by the traffic collision avoidance system (TCAS) that alerts you to other aircraft that have moved within a prescribed "safety zone" surrounding the airplane.

**Traffic collision avoidance system:** The onboard system that detects the presence of some aircraft operating in the vicinity of your own airplane. TCAS works by querying the transponders of nearby aircraft and presenting their locations and relative altitudes on a display. Alerts and warnings are issued when nearby aircraft are deemed to be a threat to safety.

**Traffic display:** A pictorial display showing any aircraft operating in your vicinity that have been detected by your traffic collision avoidance system (TCAS).

**Turn anticipation:** The function performed by the GPS computer that provides you with an advisory at the point at which you should begin your turn to the next waypoint in the programmed flight route.

**Vertical speed function:** An autopilot function that allows you to perform constant-rate climbs and descents by dialing in a vertical speed on the autopilot control panel.

**Waypoint:** A named geographical location used to define routes and terminal area procedures. Modern navigation technologies such as GPS computers are able to locate and follow courses to and from waypoints that occur anywhere in the airspace.

**Waypoint alerting:** The function performed by the GPS computer that alerts you when it senses that you are about to reach the active waypoint.

**Waypoint sequencing:** The function performed by the GPS computer that senses when you have reached the active waypoint and then automatically sequences to the next waypoint in the programmed route.

# INDEX

*All italicized page references represent figures.*

active flight plan, 22
active waypoint, 41–44. *See also* waypoint sequencing
   in autopilot targeting, 104–106
   by direct-to function, changing, 61–*63*, 69–*70*, 80–*82*
   in OBS mode, changing, 78–80, *83*
   in sequencing mode, changing, 61–*66*, 73–77
   in vector-to-final procedure, 85–86
ADF receiver, 56
*Aeronautical Information Manual*, 6
AFM, 11–12
airborne weather radar, 3–*4*, 138–148, 167
*Airborne Weather Radar: A User's Guide*, 138
airplane flight manual (AFM), 11–12
airport waypoint, 22
airspeed, 44, 119, 124, 161
airspeed indicator, 104, 164
alerts
   altitude, 108, 167
   approach mode, 57–58
   autopilot (*see* flight mode annunciator [FMA])
   GPWS, 135
   nuisance, 136, 138
   RA, 130, *132*, 170
   RAIM, 40–41, 56–58, 170
   TA, 130, *132*, 170
   TCAS, 129–134
   TCAS/VSI, 130
   terminal mode, 56–58
   top-of-descent, 54–*55*
   turn anticipation, 46–47, 171
   waypoint, 44, 46–48, 171
altitude alerter, 108, 167
altitude capture, 108–*109*, 167
altitude function, 103, 119, 167
altitude indicator, 53–54, *102*, 104, 161, 164
altitude limit control, TCAS, 130, *132*
altitude selector, 108, 167
antenna, weather radar, 142–*145*

approach
   autopilot assisted, 115–*120*
   FAF, 57–58, 85–86, 89
   GPS computer, 55–56
   IAF, 89
   localizer nonprecision, 115, *117*
   missed, 57, 91–94
   missed approach hold fix, 89
   NDB, 55–56
   in new flight plan, selecting, 27, *31*–*32*
   in OBS mode, changing, 79–80
   overlay, 55–56
   RNAV, 55–56
   in sequential mode, changing, 64
   stand-alone, 55–56
   vector-to-final feature, 83–86
   VOR, 55–56, 115
approach function, 115–*116*, 119–*120*, 167
approach mode, 57–58
approach waypoint, 22
armed functions, 111–115
arrival waypoint, 22
attitude indicator, 104–106, 164
autopilot, 3, 101–104, 162, 167
   altitude alerter/selector, 108, 167
   altitude capture, 108–*109*, 167
   approach function, 115–*120*, 167
   autothrottle, 106, 164, 168
   autotrim, 108, 168
   climbs and descents, 106–109
   disconnecting, 121
   functions, engaged vs. armed, 111–115
   heading function, 104, 110–115, 119
   ILS approach, 119–*120*
   navigation function, 103, 110–115, 119, 169
   trim commands, 106, *108*
   vertical speed function, 106–108, 119, 171
   VOR navigation, 110–115
autopilot circuit breaker, 121
autopilot control panel, *4*, *102*–104, 162–*163*, 168
autopilot master switch, *102*–103, 121, 168
autothrottle, 106, 164, 168

autotrim, 108, 168
*Aviation Weather* (FAA publication AC 00-6A), 146

Bendix/King autopilot, 6, 124–*126*
Bendix/King KLN 89B GPS computer, 13
Bendix/King KLN 94 GPS computer, *11*
  active flight plan, 22
  active waypoint, 41–44
  altitude page, 51–54
  approach, changing, 64, *68*, 78–80, *83*
  approach mode, 57–58
  approach, selecting, 27, *32*
  CDI, 42–44
  cursor mode, 19–20
  data, entering, 19–32
  departure, selecting, 27–*28*
  departure transition, selecting, 27, *29*
  descent planning, 50–54
  desired tracks, 34–*35*, 42–44
  direct-to function, 61–*63*, 80, *82*
  emergency diversions, 69–*70*
  extended pages, 18–19
  flight plan, activating stored, 22, *24*
  flight plan, entering new, 22, *26*
  flight plan, loading, 27, *30*
  flight plan page, 21
  flight plan, reviewing, 33–34
  GPS signal status, checking, 40–41
  holds, 86, *88*, 94
  IAF, 89–91
  keyboard, 14–*15*, 19–20
  leg mode, 44–*45*, 169
  manual, 6
  message, prior to IAF, 89–91
  missed approach, 89–91, 91–94
  monitor, 14–19
  navigation database, checking, 13–*14*
  navigation indicator, slaving of, 49–50
  navigation page, 42–44
  in OBS mode, 78–80, *83*
  pages, 16–19
  procedure turns, 89–91
  RAIM alerts, 40–41
  runway, selecting, 27, *29*
  simulator, 6
  stored flight plan, activating, 22, *24*
  terminal mode, 56–57
  time to arrival at active waypoint, *43*–44
  top-of-descent alert, 54–*55*
  track, *43*–44
  transition, changing, 64, *68*
  turn anticipation, 46–48
  vector-to-final feature, 83–86

waypoint alerting, 44, 46–48
waypoints, changing, 64–*66* (*see also* OBS mode)
BFG Avionics, 6
bottom-of-descent point, 50–54, 168

Casner, Stephen, web site address, 6
CDI. *See* course deviation indicator (CDI)
CDU, 160–162
certification of GPS for IFR, 10–13, 36–37
CFIT. *See* controlled flight into terrain (CFIT)
clearances, 33
climb planning, 106–109, 161
clouds, 138
cockpit automation, 3–7, 151–156, 168
cockpit weather display, 3–*4*, 138–148, 167
codes, waypoint, 74
collision avoidance system, traffic, 3–*4*, 129–134, 165, 171
color coding, of weather displays, 140
command bars, 104–106, 168
communication frequencies, 19
computer simulations, for GPS, 6
control display unit (CDU), 160–162
controlled flight into terrain (CFIT), 58, 135–136
convective wind shear, 147
course, 34–36, 168
  in autopilot targeting, 103
  in direct-to operation, 61–*63*, 80–*82*
  in final approach, 85
  in FMC CDU, 161
  on navigation page, 42–44
  in OBS mode, changing, 78–80, *83*
  in sequencing mode, 44–48
  in sequencing mode, changing, 73–77
course deviation indicator (CDI), 42–44
  in approach mode, 57–58
  in autopilot targeting, 101–103
  in holding pattern, 86–*88*
  in localizer approach with autopilot, *117*
  on navigation indicator, 49–50
  in OBS mode, 78–80, *83*, *118*
  in terminal mode, 56–57
  in vector-to-final procedure, *84*–85
crossing restriction, 50, 161
cruising altitude, 50
cursor mode, 19–20, 80, 168

departure procedure, selecting, 27–28
departure transition, selecting, 27, *29*
departure waypoint, 22
descent planning, 50–54, 106–109, 161
desired track, 34–36, 168
  in autopilot targeting, 101–103

# Index

in CDU, 161
in direct-to operation, 61–*63*, 80–*82*
final approach course as, 85
in FMC CDU, 161
on navigation page, 42–44
in OBS mode, 78–80
in sequencing mode, 44–48
in sequencing mode, changing, 73–77
direct-to function, 61–*63*
in emergency diversions, 69–*70*
in missed approaches, 94
vs. OBS mode, 80–*82*
display, GPS, 16–19, 169
distances, between waypoints, 34, 161
DVD, 7

echo, 138, 146
electronic flight instrument system (EFIS), 3–*4*, 104–106, 168
emergency diversions, 69–*70*
en route instrument monitoring, 151–156
during course changes, 94–99
of GPS, 49
during GPS approach, 58, 122
during GPS/autopilot assisted flight, 103–104, 121–124
en route procedure charts, 13, 33–34, 36
en route waypoint, 22
engaged functions, 111–115
extended pages, 18–19

FAA Form 337 11–12
final approach fix (FAF), 57–58, 85–86, 89
flight director, 3–*4*, 104–106, 164, 168
flight management computer (FMC), 160–165
flight mode annunciator (FMA), *102*, 104, 168
engaged vs. armed, 111–115
in FMC, 164–165
functions, combining, 111–115, 119
ILS approach, 119
flight plan, *10*
active, 22
loading, 27, *30*
new, entering, 19–32
page, accessing, 21
page, direct-to function, 80
page, reviewing, 33–36
page, route legs, 161
programming GPS computer with, 13–32
stored, activating, 22
flight planning, 9
GPS IFR certification, checking, 10–13, 36–37
IFR plan, filing, 36–37
weather, diverting around, 144, 146

FMA. *See* flight mode annunciator (FMA)
FMC, 160–165
fog, 138, 146

Garmin GNS 430 GPS computer, *11*
active flight plan, 22
active waypoint, 41–44
approach, changing, 64, *67*
approach mode, 57–58
approach, selecting, 27, *31*
CDI, 42–44
cursor mode, 19–20
departure, selecting, 27
departure transition, selecting, 27, *29*
descent planning, 50–54
desired tracks, 34–*35*, 42–44
direct-to function, 61–62, 80–*81*
emergency diversions, 69
extended pages, 18–19
flight plan, activating stored, 22–*23*
flight plan, entering new, 22, *25*
flight plan, loading, 27, *30*
flight plan, reviewing, 33–34
GPS signal status, checking, 40–41
ground speed, *43*–44
holds, 86–*87*, 94
IAF, response to crossing, 89–91
keyboard, 14–*15*, 19–20
manual, 6
missed approach, 89–91, 91–94
monitor, 14–19
navigation database, checking, 13
navigation indicator, slaving to, 49–50
navigation page, 42–44
in OBS mode, 78–80, *83*
pages, 16–19
procedure turns, 89–91
RAIM alerts, 40–41
runway, selecting, 27, *29*
sequence mode, 44–*45*
simulator, 6
stored flight plan, activating, 22–*23*
terminal mode, 56–57
top-of-descent alert, 54
track, *43*–44
transition, changing, 64, *67*
turn anticipation, 46–48
vector-to-final feature, 83–86
VNAV page, 51–52
VOR tracking, 111
waypoint alerting, 44, 46–48
waypoints, changing, 64, *66* (*see also* OBS mode)
Garmin GNS 530 GPS computer, 13
glide slope, 119–*120*

global positioning system (GPS), 168
go-around (GA) function, 124
GPS computer, 3–4, 6, 9, *11*, 169
  active flight plan, 22
  active waypoint, 41–44
  altitude page, 51–54
  approach, changing, 64, *67–68*
  approach mode, 57–58
  approach, selecting, 27, *31–32*
  Bendix/King KLN 89B, 13
  Bendix/King KLN 94 (*see* Bendix/King KLN 94 GPS computer)
  CDI, 42–44
  communication frequencies, 19
  course, changing, in sequencing mode, 74–77
  cursor mode, 19–20
  data, entering, 19–32
  departure, selecting, *27–28*
  departure transition, selecting, 27, *29*
  descent planning, 50–54
  desired tracks, 34–*35*, 42–44
  direct-to function, 61–*63*, 80–82
  display, 16–19, 169
  emergency diversions, 69–*70*
  extended pages, 18–19
  flight plan, activating stored, *22–24*
  flight plan, entering new, 22, *25–26*
  flight plan, loading, 27, *30*
  flight plan page, 21
  flight plan, reviewing, 33–34
  FMC, 160–165
  Garmin GNS 430 (*see* Garmin GNS 430 GPS computer)
  Garmin GNS 530, 13
  ground speed, *43*–44
  holds, 86–*88*, 94
  keyboard, 14–*15*, 19–20
  latitude of airport, 19
  leg mode, 44–*45*
  longitude of airport, 19
  manual, 6
  missed approach, 91–94
  monitor, 14–19, 169
  navigation database, checking, 13–*14*
  navigation indicator, slaving to, 49–50
  navigation page, 42–44
  in OBS mode, 78–80, *83*
  overlay approach, 55–56
  pages, 16–19
  procedure turns, 89–91
  proficiency standards, 157–158
  programming flight plan, 13–32
  RAIM alerts, 40–41
  runway names and lengths, 19
  runway, selecting, 27, *29*
  sequence mode, 44–*45*
  signal status, checking, 39–41
  simulator, 6
  stand-alone approach, 55–56
  stored flight plan, activating, *22–24*
  terminal mode, 56–58
  time to arrival at active waypoint, *43*–44
  time zone of airport, 19
  top-of-descent alert, 54–*55*
  track, *43*–44
  transition, changing, 64, *67–68*
  turn anticipation, 46–48
  vector-to-final feature, 83–86
  waypoint alerting, 44, 46–48
  waypoints, changing, 64–*66* (*see also* OBS mode)
GPWS. *See* ground proximity warning system (GPWS)
great circle route, 36, 169
ground proximity warning system (GPWS), 3–*4*, 135–138, 165, 169
ground speed, *43*–44, 135

hail, 138, 140, *142*
heading
  in autopilot targeting, 104, 110–115, *118*–119
  in GPS navigation, 44
holds, 86–*88*, 94

IAF, 89–91
IAS, 124, *126*
icing, 124, 138, 146–147
IFR certification of GPS computer, 10–13, 36–37
ILS approach, 119–*120*
indicated air speed (IAS), 124, *126*
initial approach fix (IAF), 89, 122
Insight Avionics, 6
instrument scan, 151–156
  during course changes, 94–99
  en route, 49
  during GPS approach, 58, 122
  during GPS/autopilot assisted flight, 103–104, 121–124
integrated TCAS/VSI, 130, *133*–134, 169

keyboard, GPS computer, 14–*15*

lateral navigation (LNAV), 119, 161, 164
latitude of airport, 19
leg mode, 44–*45*, 169. *See also* sequencing mode
lightning detection system, 147–*148*, 169
LNAV, 119, 161, 164
localizer nonprecision approach, 115, *117*
longitude of airport, 19

# Index

manuals, 6, 11–12
map display, 44
MCP, 162–*163*
MDA, 57–58
MFD. *See* multifunction display (MFD)
minimum descent altitude (MDA), 57–58
missed approach, 57, 89–94
missed approach hold fix, 89
mode awareness, 5, 151–156
   during course changes, 79, 94–99
   during GPS approach, 58, 122
   during GPS/autopilot assisted flight, 103–104, 121–124
mode control panel (MCP), 162–*163*
monitor, GPS computer, 14–19
multifunction display (MFD), *4*, 130–*131*, 135–*136*, 169

navigation database, 13–*14*
navigation display (ND), 162
navigation function, 103, 110–115, 119, 169
navigation indicator
   in OBS mode, 78–79, *83*
   slaving to GPS computer, 49–50, 58
   VOR radial tracking, 111–115
navigation log, *10*
navigation page, 42–*45*, 79, 162
NDB approach, 55–56
nonsequencing mode, 74, 169. *See also* OBS mode
nuisance alerts, 136, 138

OBS mode, 74, 169
   for active waypoint changing, 78–80, *83*
   for autopilot functions, 110, *118*, *120*
   for course changing, 78–80, *83*
   for holds, 86–*88*
   or sequencing mode, using, 74–94
   for procedure turns, 89–91
overlay approach, 55–56, 169

page, 16–17, 169
   altitude, 51–54
   extended, 18–19
   flight plan, 21, 41–*42*, 161
   navigation, 42–44, 79, 162
   route legs, 161, 164
   vertical navigation, 106
   VNAV, 51–52
PFD, 164
pilot operating handbook (POH), 11–12, 124
*The Pilot's Guide to the Modern Airline Cockpit*, 7, 159–*160*
POH, 11–12, 124
practice sessions, 6–7

autopilot, 127
   flight planning with GPS, 37
   following route entered in GPS, 59–60
   modifying route entered in GPS, 71, 97–99
preflight checks, 122–124
primary flight display (PFD), 164
procedure turns, 89–91
programming GPS computer, 13–32
published procedure
   missed approach, 94
   waypoint, 22

RA, 130, *132*, 170
RADAR, 138
radar attenuation, 146–*148*, 169–170
range control, TCAS, 130, *132*
receiver autonomous integrity monitoring (RAIM), 40–41, 56–57, 170
resolution advisory (RA), 130, *132*, 170
RNAV approach, 55–56
route legs page, 161, 164
runway names and lengths, 19
runway, selecting, 27, *29*

S-Tec System 55 autopilot, 6
sequencing mode, 44–48, 73–74, 170
   active waypoint, changing, 61–*66*, 74–77
   approach, changing, 64, *67*–*68*
   mode awareness during, 58
   or OBS mode, using, 74–94
   transition, changing, 64, *67*–*68*
simulators, PC-based, for GPS computers, 6
skills checklist, 127, 157–158
stand-alone approach, 55–56, 170
symbols
   terrain display, *137*
   traffic display, 130, *132*
   weather radar, 140–*142*

TA, 130, *132*, 170
TCAS. *See* traffic collision avoidance system (TCAS)
TCAS control panel, 130, *132*–*133*, 170
terminal mode, 56–58
terminal procedure charts, 13, 33
terrain and obstacle database, 135, 170
terrain display, 135, 170
terrain inhibit switch, 137–138, 170
thrust, 106, 164
thunderstorms, 144–147
tilt control, antenna weather radar, 142–*145*
time of arrival, *43*–44
time zone of airport, 19
top-of-descent point, 50–54, 106, 170

track, *43*–44
traffic advisory (TA), 130, *132*, 170
traffic collision avoidance system (TCAS), 3–*4*, 129–134, 165, 171
traffic display, 130–*132*, 134, 171
transition
   approach, changing, 64 (*see also* OBS mode)
   approach, selecting, 27, *31–32*
   departure, selecting, 27, *29*
trim commands, 106, *108*
turbulence, severe, 138, 140–*141*, 146–147
turn anticipation, 46–48, 171
turn rate indicator, 104

vector-to-final feature, 83–86
vertical navigation page, 106
vertical navigation (VNAV), 119, 161, 164
vertical speed function, 106–108, 119, 171
vertical speed indicator (VSI), 104, 130, *133*–134, 164, 169
VNAV, 119, 161, 164
VOR approach, 55–56, 115

VOR navigation, 110–115
VOR receiver, 13

waypoint, 22, 171
   adding, 64
   changing, 64–*66* (*see also* OBS mode)
   checking of, in GPS flight plan, 33–34
   codes, 74
   deleting, 64
   distances between, 34
   published procedure, 22
waypoint, active. *See* active waypoint
waypoint alerting, 44, 46–48, 171
waypoint sequencing, 48, 73–74, 161, 171. *See also* OBS mode
weather radar system, 3–*4*, 138–148, 167
wind shear, 147
winds, 44, 53

yaw damper, *126*